PUSH THE ZONE
The Good Guide to Growing Tropical Plants Beyond the Tropics

Books by David The Good

General Gardening
Start a Home-Based Plant Nursery

Florida Gardening
Totally Crazy Easy Florida Gardening
Create Your Own Florida Food Forest
Florida Survival Gardening

The Good Guides
Compost Everything
Grow or Die
Push the Zone
Free Plants for Everyone

Jack Broccoli Novels
Turned Earth
Garden Heat

PUSH THE ZONE
The Good Guide to Growing Tropical Plants Beyond the Tropics

David The Good

Push the Zone
David The Good

Copyright © 2021 by David The Good.

Cover by Andrew Chandler

All rights reserved. No part of this publication may be reproduced, distributed, or transmitted in any form without the prior permission of the publisher, except as provided by copyright law.

Good Books Publishing
goodbookspub.com

ISBN: 978-1-955289-04-7

Contents

Foreword — v

Introduction — ix

1 Twenty-two Reasons to Push the Zone — 1

2 Should You Buy a Greenhouse? — 9

3 Superpower Your Greenhouse — 15

4 Heat Islands and Shore Lines — 21

5 In Search of Microclimates — 25

6 Trees = Frost Protection — 31

7 Sprinklers and Smudge Pots, Sheets and Barrels — 37

8 Giving Your Trees the Best Chance of Survival — 47

9 Your Secret Slice of the Tropics — 65

10 Breeding for Cold Tolerance — 75

11 Strange and Marvelous Crops to Try	81
Appendix A: The Plant Hardiness Zone Map	137
Appendix B: Coconuts Outside the Tropics?	141
Endnotes	147

To Grandmom.
I have no idea how you manage
to remember so many birthdays.

Acknowledgments

This little book wouldn't exist if it weren't for a few people who started my journey to thinking differently about cold.

First I thank Craig Hepworth for sharing his fascinating greenhouse and thoughts on cold and microclimates. I also thank fellow writer Eric Toensmeier, who, though we have not met, inspired me with the tales in his book *Paradise Lot* of zone-pushing in Holyoke, Massachusetts. Through his writing I was also introduced to zone-pushing guru Dr. David Francko for the first time. I bought and devoured his book *Palms Won't Grow Here and Other Myths* which provided some solid science to back my madcap experimentation, not to mention an infectious enthusiasm for the topic. My debt to his excellent work led to my reaching out and asking the good doctor if he would write the foreword to this book, which he graciously agreed to do.

Further thanks are extended to my loyal editor and ofttimes typesetter Jeanne Logue, who has faithfully made what I write better, book after book. I thank Rick Morris, Larry Grim, and Mart Hale for always experimenting and making me think, as well as my faithful friend Allen Dovico who has likely spit-planted more loquats than anyone in history.

Extra thanks go to Sam Singleton for testing out some of my wild ideas and proving that they do indeed work. Additional thanks to Bill Hall of B & G Blueberries who shared his commercial methods of frost protection over many interesting conversations. Oliver Moore (who originally taught me to lose my fear of grafting—thank you, Oliver!) and his experimentation also must be mentioned, along with the endless encouragement of my friend and fellow plant geek Andi Houston. Joe Pierce of the Mosswood Farm Store in Micanopy was a big help with growing bananas outside their range. Paul Miller of Rainbow Star Farm in Gainesville helped me think more about the possibilities in plant genetics. Vox Day must also be thanked effusively for editing and putting my books in print and getting the two previous "Good Guide" books to hit number one in their Amazon categories—and for pushing me to get new gardening books done and out, even though he's allergic to nature and suspects I'm insane.

I also thank my parents for homeschooling me and getting me started in gardening. More regards must be given to my grandpa, Judson Greene, for getting me interested in growing pineapples and to my grandmom on the other side, Joy Perry, for being a tireless cheerleader of my garden writing. Cathy Bowers, my friend and our go-to medical expert, has also provided me with many gardening ideas over the years and inspired me with her hugelkultur projects and marvelous banana circle. Thanks also to M for inspiration and tweaking the design on my books, and to the other M for being marvelously Finnish. Andrew Chandler's illustrations make book covers great again, and I appreciate his great work ethic and deft pen. I also thank my many readers at TheSurvivalGardener.com who ask regularly, "Can I grow this here?" and keep me thinking while egging me on to great things.

Finally, I thank my witty and beautiful wife, Rachel, for being a sounding board and a cheerful companion on the garden path.

Foreword

By Dr. David A. Francko, Author of *Palms Won't Grow Here and Other Myths*

I wrote *Palms Won't Grow Here and Other Myths* in the early 2000s to help counter "zone bias" among folks living in colder winter areas. The notion that "those things won't grow here"—wonderful ornamentals like Southern magnolias and cold-hardy palms and bananas—simply because one lives on the wrong side of the Mason-Dixon line turns out to be more about cultural inertia than the actual biology of the plants themselves. At the time, we were living in Zone 6 in Oxford, Ohio just north of Cincinnati. To me, it made no *biological* sense that "Southern" ornamentals, such as crepe myrtles and Southern magnolias, were easy to find in big box stores in northern Kentucky but largely absent from the same stores located just across the Ohio River in southern Ohio.

Fast forward to 2006 when I took a new position as Dean of the Graduate School at the University of Alabama, and we relocated to Tuscaloosa. Located in West central Alabama, Tuscaloosa's long-term climate is most decidedly Zone 8. As a refugee from Zone 6, I was in plant heaven!

Yet I quickly learned that, although I had left the harsh winters behind, the sunny South had its own version of zone bias. Tuscaloosa had plentiful numbers of mature old palm trees—windmill palms (*Trachycarpus fortune*), cabbage palmettos (*Sabal palmetto*), and even much less cold-hardy jelly palms (*Butia capitata*). Many yards featured mature cycads and even some Satsuma mandarin orange trees (*Citrus unshiu*), along with the ubiquitous camellias and other keystone species of the Deep South. Yet in 2006, when I visited local nurseries to purchase these same palms and related ornamentals for my new home landscape, I could not find them. Instead, I was greeted with the same "those things won't grow here" that I heard in Ohio! Things have changed a great deal in the past decade, and zone bias is much less an issue in terms of plant availability and acceptance in most of the country—with emphasis on "most".

Many Southerners erroneously believe that "north" is all the same with regard to climate and plant adaptability, despite the fact that Cincinnati and St. Louis are more than four hundred miles south of Minneapolis. Similarly, most folks mistakenly associate the whole of the Florida peninsula with the tropics, when, in truth, the south side of Miami is some four hundred plus miles south of the Florida/Georgia/Alabama border. And so, Floridians can and do fall prey to two forms of zone bias: one, that once you hit the Florida line you can automatically grow anything anywhere, and two, that with regard to tropicals and true subtropicals, "those things won't grow here"—unless you live in extreme South Florida.

FOREWORD vii

And that's why David's new book, *Push the Zone*, is so insightful. In it, he addresses the fact that although winter temperatures in northern Florida do on occasion get cold enough to damage or even kill true tropical (and even some subtropical) ornamentals and food plants, there are numerous easy strategies for overcoming these limitations. His sections on mini greenhouses, taking advantage of favorable microclimates, mulching regimes, and the use of landscape fabric and the like offer many pathways to success for the motivated home owner to create a true tropical paradise. And because the northern Gulf Coast and coastal Southeastern U.S. share a similar climate with North Florida, his book and insights are directly applicable to folks in these areas as well. I also found many tips to be things that transcend any specific USDA zone. Now that my family and I have once again relocated back to Southwest Ohio, I am looking forward to trying them out in our new home landscape here in Lebanon, Ohio (Zone 5b).

To top it all off, David's writing style is accessible and just plain fun to read, even for the lay person with only a minimal interest in pushing the zone! Hope you enjoy it as much as I did!

> Dr. David A. Francko
> Lebanon Ohio
> October 2016

Introduction

This book won't help you grow limes in a Minnesota backyard. It won't help you grow cocoa on a Manhattan rooftop. Additionally, it won't help you find the girl of your dreams. Although you probably already knew that.

What it will do is give you a broader growing range. If you're a few hundred miles from a warmer climate where something you want to grow lives, the following ideas might just be able to get that plant working in your backyard, much to the surprise of your neighbors.

I have to start with an apology to gardeners in the far north. Though I research extensively before writing a book, I also feel that research is incomplete unless I have first pressed it into action on my own homestead. The farthest north I have lived and gardened is Middle Tennessee, so I do not have *personal* experience with shoveling snow, heating a greenhouse with propane, or getting frostbite while checking the mailbox. One disappointed Amazon reviewer of my book *Compost Everything* wrote, "I liked it ok but so much of it is about composting in a mild climate. I'm up north and a lot of it just doesn't apply."

The broader concepts do apply, however, even if the methods do not exactly fit your climate. The criticism on *Compost Everything* wasn't exactly fair, either, as most everything in that book was universal. No, I didn't cover the possibility you'd need to hack your way to the compost pile with an ice axe, or how to compost penguin carcasses, but I did cover what needs to happen for you to turn a wide range of materials into beautiful soil while avoiding throwing away potential fertility for your garden.

If I were to write this as an academic text I could certainly research exact methods for growing mangos in a heavily insulated greenhouse when outside conditions are sub-zero; however, not having had experience with that myself, I would rather not make pronouncements from thin air. My work has been done in the milder climate of the south; which, though "mild" by temperate standards, is still deadly for most equatorial plant species. When I talk about the potential for growing a papaya on the south wall of your house, move the zones up if you live in Ireland. Instead of trying for a totally tropical plant, apply the method to something not commonly grown in your climate... but something with a little more cold-hardiness than papayas (*Carica papaya*), also known as pawpaw or melon trees. Peaches (*Prunus persica*), for instance, or Japanese persimmons.

Just look to see what is growing in the zone to your south, and use the hacks in this book to see if those tempting topiaries will grow for you in your microclimate.

I grew up in the balmy tropical climate of South Florida

(USDA Zone 10). Mangoes (*Mangifera indica*) grew in my grandparents' backyard. Finger banana trees and pineapple plants (*Ananas comosus*) were scattered here and there throughout our neighborhood. Frangipani bloomed in our front planter and coconut palms hung over the million-dollar yachts that lined the canals.

But when I moved to North Florida and hit winter for the first time, I realized I was no longer in the land of hot beaches and 60-degree overnight lows in December. The first frosts hit us in November, then we had a couple of freezes (including one that dropped into the teens overnight) which made short work of the papaya, mangoes, and other tropical plants I had lovingly planted in the yard over the summer.

I know, "Cry me a river," you northerners are thinking, "It's 30 below where I live... in June!" Sorry about that. No really. I feel terrible. In fact, there's a tiny violin playing behind me.

If you live in USDA Zone 8 or south, having a simple single-layer greenhouse can get you through the winter easily. After that first winter of dead mangoes, I bought one from Amazon in order to stretch my growing zone a little.

When I set up that greenhouse for the first time, I was excited to feel how hot it got inside as soon as I zipped the doors and windows shut. I loaded it up with all my cold-sensitive plants then put a thermometer inside so I could keep an eye on the temperature.

Within the next few days, we got our first frost. The night of the frost, I closed up the greenhouse and went inside. Late at night I went out to check the thermometer. It was almost as cold inside the greenhouse as it was outside! That wasn't good. I quickly strung up a bunch of Christmas lights, plus a few flood lamps, then plugged it all in via an extension cord.

The next morning when I went outside after breakfast to open up the greenhouse, the sun was shining brightly and it was quite warm when I stepped inside. Too warm! I checked the thermometer—it was already reading in the 90s. Uh-oh. I really needed to stay on top of my venting or I was going to cook something. So I did, diligently shutting the greenhouse on chilly nights and quickly opening it again in the morning once things had warmed up.

Then I had an epiphany. I remembered reading somewhere that some gardeners in cold climates used walls of water-filled barrels to absorb the heat of the sun during the day. At night, the warmth would radiate out from the barrels and moderate the temperature in their homes or greenhouses. I decided to try it in my greenhouse. A friend of mine had a source for 55-gallon plastic drums, so I bought eight of them and placed them at regular spacings along both walls of the greenhouse, then filled them to the top with water and screwed the bungs closed. Once I did this, the temperature at night quit dropping as close to freezing... which is what I'd hoped for.

But something else happened as well. One Sunday not long after putting the drums in the greenhouse, I went to church and forgot to open the greenhouse before I left. After lunch, my family and I went to some friends' house for lunch and, half-way through the meal, I remembered the greenhouse.

"Oh no! They'll all be cooked!" I exclaimed. That got me some weird looks. Excusing ourselves, we raced home and I nervously opened the greenhouse, expected to be greeted by the odor of steamed papaya, blanched mango seedlings, and roasted canna lilies. Instead, the temperature was in the 80s. It was quite pleasant, despite the cheerfully blazing sun above.

At that point I realized: the barrels worked both ways. I no longer needed to freak out over unzipping the greenhouse during the day, and I didn't have to worry about it freezing overnight. If your freezes don't last too long, try it. You'll be amazed. When it comes to frost protection, thermal mass is your friend. The main thing I like about the water barrel method is that it's totally low-tech. I hate dealing with electricity, propane, gadgets, and wires. The simpler, the better. Plus, it uses a readily available source of free energy: the sun.

Some thin plastic sheeting and a few barrels of water allowed me to grow plants outside their range without any extra overhead.

Farther north, you'll need to be more clever with maintaining greenhouse warmth in the winter. Double-insulating, attaching it to the side of a house, creating a rocket mass heater inside... there are many ways to push it without spending too much extra money on heating; however, there will be a point of cold where you just can't pull off keeping tropical plants alive without spending some cash on a serious heating system. If I lived in a colder climate, I would likely just set up an indoor room with grow lights in the winter to keep my tropical specimens alive and skip the expense of a full greenhouse, except perhaps a simple unheated one for season extension and starting seedlings in the spring.

After having great success with moderating my little greenhouse with passive heat, I started actively experimenting with zone-pushing. I read everything I could find on microclimates. I planted seeds and transplants in unlikely locations and way outside of their natural ranges. I put orange trees next to ponds and barrels of water next to delicate fruit trees. I visited other people's gardens and noted which plants were thriving outside their supposed tolerances. I visited a pit greenhouse and saw a jackfruit growing in the North Florida town of Citra, far from the warmth of the tropics. I witnessed huge stalks of ripe bananas in Gainesville. I saw an orchid tree thriving beside an Ocala street. I planted citrus in the shade. I successfully kept a royal poinciana tree growing for multiple years without protection in my North Florida food forest. I grew pineapples next to walls and around oak trees.

In short, I tested, tweaked, experimented, observed, and zone-pushed the heck out of my plant collection... and made some surprising discoveries.

And now I'm handing over those discoveries to you.

May they bear much fruit!

David The Good

Chapter 1

Twenty-two Reasons to Push the Zone

So... why would one want to push their growing zone and fight to grow edible tropical plants out of their "natural" range? Here are twenty-two reasons!

Because You Can

Just because. That's the kind of reasoning your mom gave when you asked her why you couldn't stay up late and watch DS9, right? Turn it on the whole world!

Year-round Fruit

Yep. If you manage to pull off some of these tropical species, their fruit bearing can take you right through the holidays and beyond.

Growing Mangos

Mangos are likely the world's most delicious fruit. Just the possibility of having them in your backyard is tantalizing, is it not?

Bragging Rights

"Oh, you grow pears? That's nice. I'm growing starfruit. And coffee." Devastating! You'll be the envy of all your gardening friends. My key lime, coffee, and papayas always extracted *oohs* and *ahhs* when I gave garden tours through my decidedly non-tropical homestead.

The Challenge

Once you get good at gardening, you start to seek new thrills. Radishes and cabbage are gateway plants. Then you get into the harder stuff, the long-term addictions, like planting orchards. Eventually even that isn't enough of a thrill, so you start tying mango trees to walls and putting heat lamps and humidifiers on indoor cocoa trees.

Exotic Cooking

We grew key limes out of their natural range because my grandmother used to make the best key lime pie for Christmas every year down in Ft. Lauderdale where I grew up. Key limes were part of my culinary history, and I had to get them back. And I did. Thanks to zone-pushing my wife made Grandma's key lime recipe multiple times (though she often substituted her own grandmother's recipe instead, much to my horror. Meringue? Wrong!). Key limes are just one option. Do you enjoy mango salsa? Lemongrass tea? Peaches in Paris? Piña coladas? Getting caught in the rain? Grow your own!

Variety

Your climate only offers so many options. If you're in peach country, you might get tired of peaches. If you own a pecan orchard, you'll probably want pistachios. Zone pushing gives you a taste from another place.

Beating the Climate

Whether the globe gets hotter or cooler, the tips you'll learn in this book will give your gardens much more resiliency. When you see how things work together and how you can play at the edges of your range, you're more likely to get yields when the weather is less than friendly.

Mangos

Did I mention mangos? Because mangos are a really good reason to push the zone.

Feeding Your Soul

Many of us long for the tropics. A Caribbean beach is the wallpaper on the computer... a Hula girl on the dashboard... a palm-bedecked curtain on the window of a home office... the tropics are a symbol of freedom and long, lazy days far from spreadsheets and ice-encased windshields. A tropical garden around a backyard patio can be a place of escape from the day-to-day. A portal to peace.

Spices

Many of the world's best spices come from the tropics. What if, rather than buying all your spices, you were able to grow some of them? Cinnamon, pepper, lemongrass... marvelous!

Confusing the Experts

The experts love to tell you what you can't grow. Various "experts" told me that I couldn't grow key limes, couldn't grow papayas, couldn't grow soap nut trees... and yet, I did. And I enjoyed rubbing it in.

Being a Backyard Scientist

There is still a lot we don't know about plant genetics and their ability to handle the cold. By planting tropical plants out of their normal range and keeping records of how they respond, you can add to scientific knowledge. If you deliberately plant lots of seeds and see what happens, you may even be the originator of a new cold-hardy variety of one of your favorite fruits or vegetables.

Mangos

Mangos!

Increased Biodiversity

Adding new species to your yard makes for a more robust and complex ecosystem. Some of the tropical plants I added brought in more bees, butterflies, and other good guys. Plus, when someone asks you what the heck you think you're trying to do by planting "that tree" here, you can virtue signal and say you're "increasing biodiversity." It sounds very noble, right? And in reality, it does make for a much more interesting garden.

They Taste Good

There are some flavors you can only find in certain places. The crisp grape-like crunch of a jaboticaba fruit, tantalizing the tongue with undertones of cotton candy... the tangy refreshing juice of a fresh starfruit... the honey-and-sunshine melting flesh of a Japanese persimmon. Adding exotic and delicious plants to your garden adds excitement to the table.

They Fetch a Good Price

If you are farming for profit or considering doing so, you can up your game and your profits by selling something exotic that your customers will be amazed you grew locally. I knew a man who made as much as $1500 in a weekend by transporting one species of tropical fruit from South Florida up to North Florida and selling them to local ethnic markets. He pulled this off twice a year! What if you grew something locally, such as papaya, and sold it at a farmer's market a few hundred miles from where anyone would expect locally grown papaya to be sold? Or what if you grew cherries in apple country? Profit.

Homegrown Medicine

Moringa, lemongrass, aloe, and other medicinal plants from the tropics can be grown in your yard with proper preparation, allowing you to gain their health benefits without having to purchase them in tiny quantities for a big price.

Multiple Uses

Banana trunks can be used to feed pigs, and the leaves can be used as weed block. Moringa can help purify water. Razor-edged pineapple plants can be used as a burglar deterrent. There are multiple uses for some tropical plants, which make them worth growing, even aside from their sweet harvests.

A Hidden Calorie Stash

Someone might steal your tomatoes and watermelons. They're less likely to steal your African yams. Recognizable local crops are a target for theft; unrecognizable tropical foods will likely sail right through undetected.

Being a Rebel

Folks say "you cain't do something," an' that makes ya mad and crazy! Unleash that rebellion, and push back against the established wisdom and nature itself!

It's Fun!

Pulling off growing something unexpected is exiting. The first bloom on a lemon... the bud of a pineapple emerging from a rosette of dew-bedecked leaves... the taste of your first local nectarine... it's a great game to play and a lot of fun.

So... are you ready to get started and push the zone? Let's jump in and get growing!

Chapter 2

Should You Buy a Greenhouse?

At first glance, a greenhouse seems like a luxury for the average gardener. He imagines himself picking tomatoes in January, harvesting fresh salad greens and herbs for a New Year's party, or gathering fresh-cut flowers on Valentine's Day for the woman of his dreams (who may not have appeared yet).

But the price? Is it worth it?

You can imagine the gardener looking at greenhouses in a catalog with a calculator perched on his knee. *Hmm; how many fresh cucumbers would it take to pay this baby off... wait, what if I started all next year's transplants in it; how much would that save? Hey! What if I grew cacao and sold the fresh nibs to the organic market?*

If you twist your own arm long enough, you can talk yourself into a greenhouse, although as you're going to discover throughout this book, a greenhouse isn't completely necessary for growing plants outside their range.

Let's take a look at the "fresh greens" argument for owning a greenhouse. This is the reason many buy greenhouses. It sounds like a pretty good reason, except that in somewhat mild climates you can already grow cabbages, kale, chard, snap peas, and all kinds of good salad stuffers right through the winter. Further north, you can just eat deer. I found that growing cold-weather crops during cold weather and warm-weather crops during the summer worked well enough. Learning to store cabbages, apples, potatoes, winter squash, and other fall specialties is also cheaper than buying a greenhouse.

As for transplants, if you had a short growing season, owning a greenhouse might make more sense. I don't like transplanting vegetables in general, but I would do it if I suffered from short growing seasons. Direct seeding is a lot cheaper than owning a greenhouse; but if you do a lot of transplanting, add it to the pro column.

Here's another reason people want greenhouses: they picture stepping into the tropics on a cold day, and enjoying all the amazing plants that only grow near the equator. That's a great psychological pull. We've seen amazing indoor greenhouses in photos and thought, *Wow, what if I did that in my backyard? What if I grew coffee? Or even a little mango tree?*

That was the reason I bought mine. It served as a wondrous tropical escape. Black pepper, starfruit, kava kava, bananas, devil's trumpet, pencil trees, rare figs, coffee... I packed a lot of interesting plants into that little space.

Yet as I wrote previously, where we lived in Florida we didn't need to heat our greenhouse. Mine was just a simple plastic-wrapped steel frame with eight 55 gallon drums of water inside. Those drums moderated the temperature wonderfully and kept it from dropping below freezing even on nights where the mercury dipped into the teens. In the warmer areas of the south you can afford to manage a greenhouse like this because it rarely stays below freezing for too long. When the sun comes out, the greenhouse warms up rapidly; and the barrels never, ever freeze. This means that having a greenhouse is really pretty cheap overall, once you get past the initial purchase.

Further north, the game changes. Greenhouses get expensive. If you have the cash to burn and/or think there's a way you can profit from buying and maintaining a greenhouse, go for it. They really are fun. But they can also be a nightmare in a blizzard or a hurricane, as one friend of mine discovered when two of his greenhouses were twisted into a mess of plastic and scrap metal during a storm... along with all the orchids and other exotics he was keeping "safe" in there through the winter for his nursery business. Like a having a swimming pool, a greenhouse can be a money sink.

Greenhouses also require some maintenance and care. Plastic greenhouses tend to give up after a few years or become opaque with ultraviolet damage or algae growth. And like my friend, I almost lost mine in a windstorm some years back. Once you've spent a frantic half-hour amongst fallen and broken pots, breathlessly unclasping tie-downs in order to let the wind through as your greenhouse leaps up and down in the air, you gain a great appreciation for engineers and their clever little foundation designs. You also realize that Nature doesn't care how much money she rips from your wallet.

Every spring after the last frost date, I'd carefully remove the plastic from my greenhouse and pack it away in the barn out of the sunlight in order to stretch its life through as many years as I could. If you leave the plastic up through summer, it's going to give up a lot faster than if you take it down. Let's say the projected life of your greenhouse plastic is 3–5 years. If you're only using the cover from November through March, you could expect to get at least 6–10 years out of that plastic. It takes a little extra effort but I found it to be well worth it. When I eventually sold my greenhouse it was still sporting the original plastic cover.

You may just be better off financially not owning a greenhouse, especially once you see some of the other options I cover in this book. A greenhouse is the most obvious choice for plant protection but it's not the simplest or the cheapest.

So why did I buy my greenhouse?

The answer is simple: I'm a plant geek. I wanted to grow a starfruit tree out of its range.

In total I only got a handful of fruits. I figure they cost me roughly $250 each. The greenhouse roof was a little short for the size of a starfruit tree so I was always cutting the poor thing back in order to stunt its growth—and being in a pot really didn't help its fruit production, either.

If I had planned things differently, however, and hired a backhoe or spent a couple of weeks digging, I could have grown that starfruit into a full-size tree in a medium-sized greenhouse.

How so? Let's look at a powerful greenhouse upgrade that anyone with some time and money could pull off in his backyard.

Chapter 3

Superpower Your Greenhouse

Imagine growing mangoes, starfruit, coffee, or even a jackfruit tree in the ground in USDA zone 7, 8 or 9. These trees have almost no tolerance for the cold, yet you have them happily surviving in your suburban backyard, no pots required!

How could this be?

Thermal mass and headroom!

This possibility was opened to me when I visited the homestead of fellow plant-geek and mad scientist Craig Hepworth. I ran into him at Kanapaha Botanical Gardens in Gainesville in a bit of serendipity right as I was starting my North Florida food forest. We got talking about plants, and he told me he had his own food forest project going in the remains of an old citrus grove and I was invited to come see it any time.

I took him up on his offer and greatly enjoyed the trip—but the one thing that really stood out in my mind afterward was his greenhouse.

It wasn't anything special as greenhouses go. It was a typical hoop design, perhaps 20- ×30-foot in size, covered in inexpensive greenhouse plastic.

The difference was that the floor was dug deeper into the ground by at least four feet. It was a greenhouse over a rectangular pit—and growing from the floor of that pit was a variety of impossible things. A jackfruit, at least one starfruit, miracle fruit, and even a few small ponds with useful and edible pond plants in them. I found out from Craig that those were there for their additional heating power (our friend thermal mass!) as well as being for food production.

I don't recall how the pit walls were braced up, though I think it was with sand bags. I was too amazed by how far out of range these tropical trees were growing without needing an ounce of external energy to keep them from succumbing to the brutal frosts of mid-state North Florida.

Some of you are thinking right now, "Isn't that just a 'walipini' pit greenhouse?"

For those of you who aren't familiar with a "walipini," a walipini is an underground greenhouse with a roof that usually slopes toward the south (in the northern hemisphere) to let in the warmth of the sun during the cold of winter. Being nestled in the bosom of the earth allows the walipini to produce crops year-round even under harsh northerly conditions.

You don't really need all that much insulation in Florida, so what Craig effectually did was combine the pit of a walipini with the headspace of a hoop greenhouse. My guess is that the total headspace in that greenhouse was about 18 feet. That lets you grow smaller trees right in the ground or prune back taller trees to fit the space. A typical 9- x 12-foot greenhouse headspace isn't all that great... but when you dig a 4 or even 6-foot pit beneath it, you've suddenly created the potential for a small tropical oasis growing right into the ground, complete with fruit trees from the tropics that would never survive in your area otherwise.

You could potentially grow an entire miniature tropical food forest in a pit greenhouse with high walls and a roof...all in a little backyard. Further north, you might pull off peaches, cherries, plums, apricots, Japanese persimmons, and loquats hundreds of miles north of their range. If planned out well, you could be harvesting edibles year-round.

What About a Tiny Food Forest... in a Greenhouse?

As I explain in my little book *Create Your Own Florida Food Forest*, a food forest (or "forest garden") is comprised of a series of layers just like a natural forest. I didn't come up with the idea, of course—it's based on the work of visionary gardener Robert Hart. The top layer is the canopy, followed by the sub canopy comprised of shorter trees, then the shrub layer, the herbaceous layer, and then the ground cover layer. There's also an optional vine layer, and other folks also count the root layer and even a fungal layer or perhaps an aquatic layer where one might have a pond or a stream running through the food forest.

The concept is to create something that pretty much takes care of itself, like a natural forest does. A forest has every piece filled with something, meaning where weeds aren't a problem, bare soil isn't an issue, and fertilizing takes place due to the action of falling leaves, rain, and the working of fungi, animals, and myriad tiny creatures—both macro and microscopic.

A greenhouse isn't the ideal way to create a system like this, but you may be surprised by how much you can grow in one and how dense and productive the system can become.

It's probably best to start with shorter trees for the canopy, not massive trees like kapok, jackfruit (more on the potential for jackfruit later), Brazil nut, or full-sized mangos; that would be trouble! I'm also going to combine the sub canopy with the shrub layer and the ground cover layer with the herbaceous layer since we don't really have a lot of headspace.

Here's what I would plan out in the South:

Canopy:

Star Fruit, Guava, Acerola cherry, Dwarf Cavendish banana, Moringa, Patio/Condo Mango, Cashew, Dwarf June Plum, Dwarf coconut palm, Jaboticaba, Lime, Lemon

Sub canopy/Shrub Layer:

Katuk, Tea, Pigeon peas, Cocoplum, Coffee, Surinam cherry, Cherry of the Rio Grande, Chaya, Miracle Fruit, Monstera, Natal plum

Herbaceous/Groundcover Layer:

Gingers (various), Kava kava, Root Beer plant (careful—it'll take over!), Longevity spinach, Okinawa Spinach, Waterleaf (*Talinum fruticosum*), Cuban oregano, Basil, Hot peppers, Everglades tomato

Vine Layer:

Vanilla orchid, Black pepper, Yams (spp.), possibly including edible *Dioscorea bulbifera*, Beans, Jicama, Perennial cucumber (*Coccinea grandis*)

Imagine fitting that group of amazing tropical edibles into your backyard! You could pull off a greenhouse like this for $3000 or less, depending on how much digging you're going to do. Heck, it would be worth it just for the bragging rights.

Farther north you might plant a food forest of plants that can take some cold but not terrible amounts. Trees such as peaches, nectarines, sweet cherries, plums, apricots, almonds, thornless blackberries, rabbit-eye blueberries, grapes, etc. The design of the greenhouse would have to be more robust to ward off heavy snows that could collapse the roof and some supplementary heating might be required for long periods of cold.

The biggest challenge once you plant a system like this will be keeping the more aggressive trees from shooting through the top of the greenhouse. You need to prune judiciously. Though most of these trees are smallish, some will grow right through the roof. Cut out the central leaders of the trees when they near the roof, bend branches, and tie them down to cinderblocks as needed. Be sure to get everything well under control before the roof needs to be constantly sealed during the cold of winter.

Now you may rightly ask, "David The Good... did you do this yourself?" The answer, sadly, is no. Not completely. I visited some working greenhouse systems with in-ground tropical trees, plus I grew a lot of tropical plants in my ground-level greenhouse via pots, but I never made it to the point of growing a complete mini-food forest in a greenhouse.

Time and money got away from me... and then I sold my property and moved to the tropics, leaving you all behind to pick up where I left off.

It will work, though. Where there's a will, there's a super-charged pit greenhouse.

Chapter 4

Heat Islands and Shore Lines

You may be longing to move to the country to start farming and gardening seriously, but before you do—did you realize that there are some climate benefits to living in an urban location?

According to the U.S. Environmental Protection Agency, "The term "heat island" describes built up areas that are hotter than nearby rural areas. The annual mean air temperature of a city with 1 million people or more can be 1.8–5.4°F (1–3°C) warmer than its surroundings. In the evening, the difference can be as high as 22°F (12°C). Heat islands can affect communities by increasing summertime peak energy demand, air conditioning costs, air pollution and greenhouse gas emissions, heat-related illness and mortality, and water quality."[1]

Sure, the EPA puts a negative spin on it—big surprise—but that heat island effect is your friend when it comes to growing tropical plants outside the tropics.

I've seen trees growing in the city of Ocala (North Central Florida) that would die outside of that urban microclimate. A few coconut palm trees are growing miles from their normal range inside the shelter of Tampa Bay (West Central Florida). I've seen a gigantic fig tree tucked inside the confines of an apartment complex in a freezing area, safely outgrowing the common shrub-like habit of figs pruned by frost. Concrete, walls, pavement, automobiles, energy usage and people combine to raise the temperature of a city above the surrounding countryside.

Another case in point: my friend Jo has a purple orchid tree growing in her front yard inside the city of Ocala. That same tree would have frozen to the ground on my old property twenty minutes outside the city. Its location between buildings and over a sidewalk keeps it just warm enough to stay happy and blooming despite the non-tropical latitude.

The Warmth of the Ocean

If you live near the coast, consider yourself blessed. The climate of Jacksonville (Northeastern Florida) is comparable to that of Groveland, despite the former being located farther north. This is why there are coconut palms growing inside Tampa bay but not in Orlando.

Like the barrels of water in my greenhouse, the ocean functions as a huge repository of warmth on chilly nights. The farther you get away from the ocean, the worse the overnight lows become—and the hotter the summer highs. The center of the state of Florida is a ridge with rolling hills and little to moderate the heat of summer or the cold of winter. In my previous location south of Gainesville, I could drive an hour east or an hour west and see a lot more tropical foliage growing than would survive on my homestead.

More than one person in the Ocala area told me "I could never live in South Florida—it would be way too hot." I'd just laugh. I grew up in Ft. Lauderdale and I can tell you from experience: it never, ever gets as hot there as it does during the middle of a Central Florida summer. The ocean keeps it both cooler and warmer. Where I currently live in the tropics is even milder. Temperatures normally range between 74 and 87 degrees, even though you'd think its equatorial location would lead to sweltering misery.

If you're not tied to a particular location, you have a few options in finding a place where you can grow warmer climate plants outside with little or no protection. You can move southwards towards the equator... you can move into a warm, urban area... or you can move towards the coast. Heck, you could combine all three and move to Miami; however, the crime rate down there takes some of the fun out of gardening. There's nothing that dampens horticultural enthusiasm like having some hoodlums steal all your mangoes or break into your tool shed three times in one week. No fun.

Getting close to the ocean or into the city is more expensive than living in the country, unfortunately, which is part of why we moved into the middle of my home state. Land is abundant and the soil can be pretty decent by Florida standards, depending on where you settle.

If you already live in the city, I recommend trying some trees and plants from farther south. Patios and pool areas are excellent for planting small fruit trees. If you're in an apartment, try growing some tropical plants in large pots on your deck or garden area.

I've seen queen palms growing in North Florida between an apartment wall and a pool. If those palms were planted out in the open, they would be toasted by frost. By the building, in that urban heat sink along with the additional thermal mass of a big swimming pool they looked as happy as if they were in Tahiti.

The way to find out what works is to plant a lot of things and see how they do in your area. I'll bet there are places in your yard right now that are warmer because of their location.

I once visited a friend about seventy miles east of my old house in Marion County. He lives in Ormond Beach, right near the ocean, which was actually a little north of me. In his neighborhood people were growing royal poinciana trees and sea grapes—both decidedly tropical species. The ocean made all the difference.

If you get to choose, choose warm areas. If not, it's time to start seeing what you can do around your existing property... and that's what we're going to focus on in the next chapter.

Chapter 5

In Search of Microclimates

The first couple of winters at my house in North Florida, I spent a lot of time finding the microclimates in my yard. Quite a bit of this happened on accident.

I planted mango seedlings here and there around the yard as well as a half-dozen small papaya trees. I also planted cassava, guavas, pigeon peas, and other tropical plants.

The first freeze didn't kill them all, but it did kill some of them. You could tell which areas were warmer in the yard by the relative level of frost damage on the plants.

There are easier and cheaper ways to determine microclimates, of course. You could simply plant a warm-season vegetable like bush beans late in the year and see which ones are killed when the frosts arrive; and which ones, if any, live.

Simpler still, you could watch the weeds in your yard. If you're like me and not much of a lawn person, there are plenty of weeds mixed in with your grass. One of the best for determining microclimates in the Deep South is the pesky shepherd's needle plant, known more properly as *Bidens alba*. I had those growing all over my yard in every corner of my patio area, food forest and garden beds. They self-seed prolifically and are a favorite of the bees. They're also edible. Those benefits aside, where they're most useful to you as a microclimate hunter is in their lack of tolerance for cold. If you get a frost, they brown right up. If you don't, they'll live through the winter and keep flowering in the spring. What I found quite interesting was how we'd get a night below 32°F and there would be patches of shepherd's needle that wilted into limb huddles of brown... and patches that stayed upright and unscathed, often within mere feet of each other.

Noticing these patterns is important. I daresay most of us would miss the clues these weeds reveal about the warmer and cooler areas in our yards.

Back to my mangos and papaya. A few papayas were kept in pots on my south-facing porch, and those survived despite being out in the freezing night air. Nary a sign of damage. The ones out in the yard were all killed to the ground, thanks to a night in the teens. So much for my dreams of a tropical paradise!

The papaya from the porch were planted along the south wall of my house in the spring. They reached about 10 feet tall and bore me some fruit before the chill stopped their growth; and the next year's frost lopped off their tops but didn't kill the trees completely. In the spring they came back from their bases and at least one of them bore a few more fruit, but nothing like the first year against the wall.

Undeterred, I planted more papayas against the wall but even closer to the concrete. Those bore quite a few large fruit before their icy decapitation the next winter. What I really needed was a shorter papaya. The wall was warm and they were sheltered in large part by the overhang of the roof... until they reached past the edge of the overhang and were exposed to the open night sky and the full brunt of descending cold.

I'll come back to the power of a south-facing wall in a subsequent chapter, as multiple years of experimentation eventually proved it to be the best place in my entire yard to grow tropical plants.

There are different ways to "zone push" and find and/or create microclimates to grow things where they won't normally survive. Rocks and water are used by Sepp Holzer (if you don't know who he is, look him up and buy his books—the guy is brilliant) to grow cold-sensitive crops in the Alps. In a place where there really should be nothing but super cold-hardy pines, he's growing cherries, apples, and apparently even lemons (though he won't tell anyone how... darn you, rebel farmer!). The rocks and water hold heat and moderate harsh temperatures thanks to their thermal mass. If you had a south-facing rock wall with some pockets of soil in it for plants, you could likely do some pretty good zone-pushing.

A few years ago I created two passive heaters for one of my small guava trees: a pair of large gin bottles filled with water and painted black. I placed them, along with a ring of stones, beneath the two-foot canopy of the recently-planted tree, hoping to keep it just a little warmer during the upcoming freezes. The poor thing froze to the ground anyhow. Maybe they were sad I hadn't given them the gin? The coldness of the vacuum above is hard to beat. Nothing kills plants like a wide-open sky on a freezing night. The heat is sucked away into the void, bottles notwithstanding. There just wasn't enough thermal mass—and that was the night the temperature hit 12 degrees. Fortunately, the guava came back from the ground the next spring. When it did, I transplanted it to a spot along the south wall by the back door of the house where it thrived and never froze again.

I had a plan at one point to create a big pond. Once completed, I wanted to build a floating raft in the middle with a mound of soil on top. On that mound of soil, I was going to plant papaya trees. The island would be over the awesome thermal mass of all those many gallons of water. Would it work? I leave that for you to try. I started digging the giant pond, but my wife objected since she was certain our two-year-old would drown in it. I decided that was a good point and ceased construction. In my mind, though, it was a beautiful pond. And the papaya in my mind—wow, you should've seen 'em!

I did end up using one pond as a way to protect a blood orange tree from the cold. It wasn't a proper pond with a liner in the ground though. It was a hot tub I found by the side of the road that I patched up so it was watertight. I dug it halfway into the ground, threw some dirt in the bottom, and planted it with native flag irises (which I initially thought were cattails), Chinese water chestnuts, and various floating pond plants I found here and there. To top it off I added a couple of goldfish from the pet store and some mosquitofish to eat mosquito larvae.

Directly beside that half-buried hot tub pond, I planted a beautiful little blood orange tree. Though I did throw sheets over it on some really cold nights, on other nights I didn't and it seemed to do fine either way. The radiant heat of that water was a great thing and the tree grew faster than most of my citrus trees. Not getting damaged in the winter helps plenty!

While we're on the topic of citrus, I'll end this chapter with the tale of two trees and one freezing night.

When we first moved into our North Florida property and before I discovered just how bad the greening virus had become in our state (which led to my later recommendations *not* to plant citrus anywhere in Florida!), I wanted a bunch of citrus trees in the yard. I planted three oranges, a couple of grapefruit, three kumquats, a calamondin, and another grapefruit that turned out actually to be a tangerine that had been mislabeled.

A funny thing happened the first frost we got that year. I didn't think the temperatures were going to dip below freezing so I didn't bother covering anything.

By the afternoon of the next day, I realized that had been a mistake. One of the orange trees had frozen severely, to the point where its trunk had actually split in multiple places. Yet, just twenty feet away from it, another orange tree was fine. It was slightly touched by the frost, but the damage was negligible.

How could this be, I thought, *what could be the difference?* Then I looked up.

Chapter 6

Trees = Frost Protection

If you grow your plants out in the open, freezing nights can be deadly.

The Gainesville/Ocala area where I lived is marginal for citrus. It has nights that reach down into the teens. I know, some of you northerners laugh at that (I'm looking at you, Minnesota) and wish you had it so good, but the teens can be deadly for citrus.

When I planted the two little orange trees I mentioned in the last chapter, one was totally out in the open, and the other one was partly under the canopy of a large oak tree.

One tree froze to the ground yet the other suffered only minor frost burn.

Guess which tree did what? You guessed it: the one in the open kicked the bucket.

My friend Craig with the cool greenhouse once told me that he'd been growing pineapple in his greenhouse for quite a while. He lives at the edges of Zone 8 and 9; a place where pineapples are allegedly unable to grow, so he'd assumed they needed the protection on freezing nights.

The problem with his pineapples was that they were a pain to weed. If you've ever dealt with pineapple plants before, you know how nasty their serrated sword-like leaves can be. After growing and harvesting them for a time, he decided to clear out a bunch to make room for less painful plants in his greenhouse. Craig told me that he simply ripped out the plants and tossed them around the trunks of some oak trees to deal with later. To his surprise, they grew and continued to bear pineapples despite freezing nights.

After hearing his story, I decided to do the same thing. I'd been growing pineapples in pots for a while and hauling them in and out of my greenhouse.

The pineapples sailed through multiple winters without a hitch, so I also added strawberry guava (*Psidium littorale)*, naranjilla (*Solanum quitoense*), and edible hibiscus plants beneath my oak trees.

At one point I thought the oaks were just a pain-in-the-neck tree that dropped limbs and took up space where I could be growing edible trees. After experiencing their frost-defying power, I completely changed my opinion.

How Trees Provide Frost Protection

Part of the way trees protect your plants from freezing is through capturing still, warm air on a cold night. Their canopies, even when leafless, hold back the cold of space above.

TREES = FROST PROTECTION

Higher numbers of trees make an even bigger difference, sometimes having double-digit effects on temperature.

Tender plants in the middle of a lawn? Doomed. Tender plants under a bit of forest canopy? Better.

Another way trees protect the plants around them: thermal mass.

How's that?

Well, imagine a tree trunk as a column of water held in place by cellulose fibers. Water takes some serious energy to freeze and even in the winter trees are able to keep their sap from freezing through.

I've used barrels of water to keep my greenhouse from freezing; trees can have the same effect on plants near their bases.

Sometimes there are solutions to gardening problems that don't require serious infrastructure. Knowing what I now know about trees and their ability to provide frost protection, I'm always horrified to see new homesteaders start their properties off by clearing the forest.

Work inside it and clear minimally or you may be giving yourself a harsher climate than you'd like.

Look for a tree, grab yourself a couple of plants from further south, pop them in by the trunk if they can take full shade... or plant them further out and nearer to the edge of the oak canopy if they can't, then see how they do through the winter.

Since avocados grow easily from avocado pits, I spent some time doing a multi-year experiment with planting them here and there around my yard, in and out of the tree canopy. Actually, I didn't even plant some of them on purpose—they just came up in some of the "melon pits" I created with kitchen scraps and manure (more on that excellent method of composting can be found in my book *Compost Everything: The Good Guide to Extreme Composting*). Throw out avocado pits and sometimes you get avocado trees.

These were not cold-hardy avocado cultivars. They were just avocados from the supermarket or from family down south, grown in far-off tropical climes like "Mexico," "California" and "Fort Lauderdale."

What happened was exactly what I'd expect to see after witnessing the way orange trees reacted to shade or open sky on freezing nights: the avocados in the most open areas were damaged by frost whereas the avocados in full shade beneath the oaks were completely untouched.

Many gardening texts will tell you that various fruit trees need full sun to fruit. This is unfortunately true for many, but not all fruit trees. I've seen oranges and other citrus growing well and bearing fruit in full shade. I've even seen an orange grove interplanted beneath towering live oak trees. One time I was helping paint a house and beside the back patio, in almost full shade, there was a nice little orange tree loaded with fruit. Japanese persimmons, on the other hand, have stubbornly refused to even bloom for me in full shade.

Since most of the agricultural extension publications you read online are predominantly concerned with the commercial growing of fruit and nut trees inside their "proper" range, it's sometimes hard to find out which trees will fruit in the shade and which ones will not.

That's where experimentation comes in. Plant fruit trees in full shade and half shade and see what happens. How much money will you be out per tree if it doesn't work? $25? And if it does work, how much organic fruit does it take to reach $25? Not a whole lot. If you start from seeds, like I did with avocados, you have no money invested in the experiment at all. I've also started plants from cuttings and stuck them here and there to see what would happen. One success story was the Natal plum (*Carissa macrocarpa*) I placed beneath a calamondin tree as a little 6-inch tall start. It sailed through some freezing nights with only a touch of damage despite its reputation as a zone 10 fruit.

Big canopy trees such as oaks, hickories and even pines are some of your best allies in the fight against cold. They moderate the temperature and provide shelter to species that would otherwise perish on an icy night. Embrace the canopy and you'll grow plants you never thought possible.

Chapter 7

Sprinklers and Smudge Pots, Sheets and Barrels

Before we go any further, let's cover three simple methods you can use to gain a few degrees on a cold night. Those few degrees are often the difference between minor damage and a dead tree. We'll start with sprinklers.

Sprinklers

On February 13, 2011 we suffered a very, very cold night when my blueberries were in full bloom. I really wanted fruit and I wasn't willing to accept the loss of all that potential. I remembered that a friend of mine ran a blueberry U-Pick so I gave him a call.

> ***DAVID THE GOOD:*** *"Bill, what are you doing to keep the blueberry blooms from freezing tonight?"*

BILL THE BLUEBERRY GUY: *"I'll be running the irrigation all night," he told me, "they'll be fine."*

DAVID THE GOOD: *"Wait... you're running the sprinklers? And that works?"*

BILL THE BLUEBERRY GUY: *"Yeah, buddy!"*

DAVID THE GOOD: *"Don't they freeze into blocks of ice?"*

BILL THE BLUEBERRY GUY: *"Sure they do, and you just keep running it. They'll thaw out and keep right on going."*

This sounded crazy. My friend, my good friend, was insane.

Still, I had to try it just in case, so I put a sprinkler out and ran it for the entire night on my blooming blueberries.

The next morning, I came out to find them completely covered in icicles with the sprinkler still faithfully oscillating back and forth. The ice was as thick as an inch on some branches, much to the entertainment of my children who thought the whole thing was rather hilarious. I admit—it did look strange. The blueberries resembled small lucite-dipped weeping willows. Following Bill's advice, we kept the sprinkler running until all the ice melted sometime in the middle of the day. Despite my fears, the blooms continued to develop normally into delicious homegrown blueberries.

As I dug further into commercial frost protection techniques, I discovered that using sprinklers for frost protection is a common trick used by professional growers to save blooms and young fruit from frost. Remember—I grew up in the tropics, so I wasn't an old hand at frost protection techniques. You'd think it would totally wreck the plants to encase them in ice but it doesn't. In fact, it keeps the plants hovering right around freezing without letting them freeze all the way through, thanks to the slight heat caused by the phase transition from liquid to solid.

I went on to use this method to protect future blueberry crops as well as on my peaches and mulberries.

Overhead irrigation has its downsides in that it uses a lot of water and also requires running hoses and sprinklers around your yard or putting in pipes. It's also not easy to use on larger trees without some awesome water pressure.

Smudge Pots

For those of you who have never heard of a "smudge pot," count yourself (and especially your lungs) lucky. As the *Encyclopedia Britannica* defines it:

Smudge pot, device, usually an oil container with some crude oil burning in the bottom, used in fruit orchards, especially citrus groves, to provide protection against frost. The smoke serves as a blanket to reduce heat losses due to outgoing radiation. Because of the air pollution they generate smudge pots have been generally supplanted by other means of frost protection, such as smokeless burners using natural gas.[2]

Some of the old-time citrus growers still remember the "smudge pots" that generated large clouds of smog in the orange groves on frosty nights in the hope of keeping the coldness of a clear night sky at bay.

As one man who grew up in the Southern California citrus industry reminisces:

While some thrifty growers still burned stacks of old tires, the majority turned to the coke-burning units ("smudge pots") as they became available and later invested in the more efficient oil-fueled heaters. All methods of orchard heating are costly, not only in the price of the fuel consumed, but in the wages paid the manpower required for the operation and maintenance of the system. I recall that as a general figure it was estimated that fuel oil costs accumulated at a minimum of $50-per-hour (1939 dollars!) for every ten acres heated. The grower, his family, and perhaps some hired labor (usually at about 50 cents/hour) would be involved in each firing of the heaters. During World War II, the Citrus Association arranged for Mexican laborers ("Braceros") to work in the groves, and often made them available for heater firing and maintenance. As to the heaters, the early commercial models were simple sheet metal stacks with a grate and air adjustment holes in the bottom. The fuel of choice was a mixture of coke briquettes and oil-soaked wood, purchased in large burlap bags, which was lighted from a hand-held torch containing a mixture of oil and gasoline. In *theory, the lighted heater could be shut down by placing a cap on the stack. This was seldom effective and a lighted heater was usually allowed to burn itself out. The next day, the large bags of the coke/wood mixture, were stacked on sleds or stone-boats, which were dragged by tractor or horses through the groves and stopped to fill each heater in preparation for the next night.[3]*

In an article from *The Harvard Crimson*, it's stated:

The idea behind smudging is ridiculously simple. Squatting in the middle of orange groves all over California are filthy black smudge pots. The pots have a ten-gallon belly at the bottom and a four-foot smokestack rising out of the belly. There's a hole on the top of the belly where cheap and grimy smudging oil is poured in, and there's a hole at the top of the smokestack where grimy smudging smoke belches out. When the pots are lit and roaring, they produce an astounding amount of smoke, some noise, and a little heat.[4]

On cold nights with minimal cloud cover, the warmth of the ground radiates away into space and temperatures drop rapidly... unless you fill the air with clouds of smoke and help keep the heat of the ground from dissipating as quickly.

The idea of a home gardener lighting multiple smoke-belching pots in his front yard orchard is quite entertaining to me. You should totally do it. You can still occasionally find smudge pots in antique stores around Florida but I've never tried this method of frost protection personally.

The closest I ever got was one frosty spring night in Tennessee when I tried to protect the blooms on one of my pear trees by lighting my barbecue grill and stuffing it full of rocks, then rolling it beneath the branches in the idea that perhaps it would provide enough radiant heat to stave off the frost.

The blooms still froze. And I broke my barbecue grill, which my wife still brings up occasionally. "Hey... how about we grill these steaks, David? Oh, wait! Someone broke the grill!"

Women are cruel creatures with exceptionally long memories.

Clouds of smoke will gain you a few degrees on a cold night but your neighbors might not appreciate your smudge pots as much as the trees do.

Sheets

Sheets or "frost blankets" are one of the most common methods of frost protection you'll see in Florida. When trees are young they're usually much more susceptible to frost than they are when they grow older. Throwing a blanket over a young orange tree on a 25-degree night makes the difference between future oranges and a dead tree. You must protect young tropical trees until they grow enough to withstand the cold on their own.

Sometimes they'll never be able to handle it, unfortunately. I visited a Master Gardener in Groveland and saw a big wooden frame in his yard with a 12-foot tall dead tree inside it.

I asked what it was and he told me. "That was my lychee tree. I built a frame around it and put plastic over it during the winter, like a mini-greenhouse. It grew really well for the first few years, and I always protected it. This last fall it was so big I figured it could probably take the cold. I was wrong."

Oranges gain cold tolerance to levels below freezing as they grow older, whereas some species, like mangos (and maybe lychee) really can't take much frosty weather no matter how big they grow.

Back to sheets.

The idea is to keep some of the warmth of the soil around your tree. The ground is warmer than the sky above, and though trees don't generate their own heat like we warm-blooded gardeners do, preserving what little they have with a blanket really helps.

Though there are frost blankets you can buy which are specifically made for covering plants and trees, I've had great luck buying second-hand sheets and blankets from local thrift stores and using those to keep my trees warm or cold nights. Grab some clothespins and bundle of blankets, and you can cover a dozen or more trees in an hour or so. The best clothespins for this task, incidentally, are the very sturdy American-made pins sold by Herrick Kimball. I own multiple sets.

They're expensive compared to cheap Chinese clothespins but they bite like a pit bull and last for ages. You can find them at classicamericanclothespins.blogspot.com/. As for the blankets, I spent an average of $2 per each. My wife also sewed some of the blankets and sheets together to make super-size frost protection covers for our taller tender trees. I would wrap them up and around the trees on cold nights. This made it look like there were terrifying spectral creatures gathered in our yard. I walked out one evening after covering the trees during the day and almost jumped when I caught multiple tall, robed figures looming out of the dusk in the corner of my eye.

One thing you don't want to use if you can help it: plastic. Anywhere it touches the leaves, they'll usually be burned by frost. My guess is that it transfers the cold too easily compared to the insulating power of cloth. Plastic will also cook your trees if you fail to uncover them the next morning. When I cover with sheets and blankets, I can leave them on for multiple days without harming the trees. I've had blanket-shrouded oranges standing in my yard for most of a week before since I really didn't want to remove and replace their covers over multiple nights of frost. It doesn't hurt the trees.

When you wrap your trees in sheets, it seems to work best if the covers touch the ground completely. I used to always keep rocks and bricks around my trees so I could throw blankets over them and then pin them to the ground around the drip line of the tree. You want the warmth of the soil to head upwards overnight... and the freezing air to stay out.

Barrels

For extra-special trees you can keep some 55-gallon drums ready to provide warmth on cold nights. The day before the first forecasted frost, place the drums right next to the trunks of the trees and fill them up with water. Congratulations! You now have a low-tech space heater!

I planted a seedling loquat in my front yard food forest about six years ago. After five years, it bloomed in the late winter (as loquats do, darn it) and was covered with plenty of tiny fruit. I was excited to taste them for the first time and we were almost out of the clear... when we got some bad news. A hard freeze was on the way! Loquat fruit are only cold-hardy down to 25 degrees, though the rest of the tree can easily handle the teens, so protecting loquat fruit from frost is vital.

The tree was way too big to cover with blankets at that point and I didn't have a massive parachute handy. Thinking fast, I decided to put a barrel of water under one corner of the bloom-laden tree, figuring I could at least protect those fruit and have some to taste. I also threw a blanket over those branches above the barrel to hold in some of the rising heat overnight.

The thermal mass of that water, along with a few thrift store sheets and blankets made a big difference.

The only place on the entire loquat tree that held fruit was the portion right around the barrel of water.

Water holds a lot of energy. It works in a greenhouse, and it works in your yard. You can use this trick to protect any number of tender plants. For the best results, tuck a barrel right next to the trunk and cover the rest of the tree with blankets or sheets to keep the heat of the water in overnight.

Now let's look at some ways we can manage the trees directly to increase their chances of winter survival.

Chapter 8

Giving Your Trees the Best Chance of Survival

Imagine two men.

One of them is trim, plays on a semi-pro soccer team, eats a nutrient-dense diet, lifts weights and drinks and smokes only occasionally with friends.

The other man smokes two packs of Kools a day, eats donuts or pastries for breakfast, sweats when he walks up a single flight of stairs, eats fast food for lunch every day and drinks a six-pack at night while eating microwaveable burritos and ice cream in front of the television.

Which of these two men do you think would do better on a three-day hike through a portion of the Appalachian Trail? Or on a canoe trip? Or on a blind date?

The difference is easy to see. We know how we should treat our bodies, whether or not we do so. Good health leads to success in multiple areas—poor health leads to failure.

If we pay attention to the health and care of our trees, they'll do better as well. We usually don't think about trees very much, other than to occasionally water or throw some 10-10-10 around their trunks... but if you want to grow tender trees outside of their natural range, you can increase their odds with a little care.

Pruning

I used to have an aunt who constantly pruned and shaped her trees. In general there's nothing wrong with that. Trees respond well to shaping and can be tweaked into some remarkably interesting forms as topiary enthusiasts can attest. The problem is that pruning in certain ways reduces a tree's ability to survive frost.

Consider a citrus tree. Its natural shape is like a big umbrella. This growth pattern is great for capturing heat on a cold night. The rising warmth of the ground is held in. Though the top of the tree will suffer frost damage, the trunk, the branches and most of the leaves beneath will often sail through a frost without damage.

If you prune up your citrus tree to "open it up," you also open it up to the cold of a winter night. It might be nice aesthetically, but it's stupid practically.

Another bad idea is to prune trees late in the year, particularly tender trees. When you prune a tree, you stimulate new growth. New growth is much more subject to frost damage than older growth that has "hardened off." A tree pruned in

October may burst into a flurry of growth in November, then be destroyed by cold in December. Infection can set in from damaged areas, and you may lose a tree that otherwise would sail through the cold.

Don't open up the tree canopy and don't prune late. Just remembering those two rules of thumb will help immensely.

Now when a frost does hit and burns your tree, do not get out the pruning shears! Let the tree sit there until all danger of frost has passed, brown leaves, twigs and all. I repeat: do NOT "clean it up." Aesthetics are not important—the tree's survival is. If you prune off the damage, you are removing a layer of protection from the tree and may also be encouraging new growth at the same time. Let it sit, even though it's ugly and sad. Those dead leaves and branches are still providing some protection for the living wood further in.

Mulching

When I lived in Tennessee, I was always told that the secret of keeping things alive through winter is mulch, mulch, and more mulch.

The same idea is often espoused by gardeners farther south—yet in a mild climate mulching may actually *reduce* a tree's ability to survive a frost.

"But David, doesn't mulch keep the ground warm overnight and protect the roots?" I can hear you asking right now, "everyone knows that mulch stops the ground from freezing!"

True... but what if the ground rarely freezes in your area?

Remember what I've shared already about thermal mass? Rocks, walls, barrels of water...

...well, right beneath your tree is an abundant source of overnight warmth: the earth itself.

Mulch or ground cover acts as a blanket over the soil, keeping the warmth in overnight. That's great if you live farther north and are trying to protect the roots of dieback perennials like Echinacea or lemon balm... but it's lousy if you're trying to keep a tree alive through a few-hour dip below zero. Mild climate frosts usually happen quickly and are gone the next morning a little while after the sun rises. The ground doesn't really get a chance to freeze here.

Have you ever visited a citrus grove and wondered why the trees are surrounded by bare soil? This is one of the reasons. That bare ground is radiating heat upwards all night long.

If you mulch, you trap in the warmth. If you clear the ground, you let the warmth out.

If we were to have a day or more of below-freezing weather, bare ground would be in danger of freezing... but if that happened, we'd all have dead trees anyhow, so why worry about it?

A way to increase the thermal mass of the soil as well as the transfer of heat upward to your trees is to water the ground well before a freeze. I know, you're thinking, "the last thing I'd want is to be soaking wet on a cold night," but remember: trees are cold-blooded (cold-sapped?). Get that ground soaking wet, and you add more warmth than can be carried in dry sand.

Mulching is a great thing in your gardens and it's good for your trees most of the year. On a cold night in a warm climate, however counterintuitive it sounds, mulching may hurt more than it helps.

Now if you do live farther north and the ground freezes regularly in the winter, mulch, mulch away. That blanket of decaying wood chips should hold plenty of air and moisture that will ward off the worst of the weather and may make the difference between life and death for your more tender specimens.

Keeping Your Trees (Less) Chill the Chilean Way

At one point I seriously considered moving to Chile. It's an incredible nation, stretching from near Antarctica in the south, through the lush temperate rainforests in the Southern Cone, onward into the Mediterranean climate of Santiago and then into the bone-dry Atacama desert. To the west is the rolling surf of the Pacific Ocean, to the east are the looming Andes Mountains.

One of the places where Chile really shines is in its farms and agriculture. Vineyards and orchards cover the landscape in a patchwork of varying green in the north... but this isn't a travel brochure, so I'm going to stop there. For the scope of this book I'll just cover the way many avocados are grown in Chile.

Though there are cold-hardy avocados from Mexico now available in the states, they aren't the cultivars we see commercially. Those are tropical species usually grown only in South Florida, southern California, and in nations through Central and South America.

Chile, for the most part, is not located in the tropics. The Atacama Desert is closer to the equator but the almost complete lack of rainfall limits its agricultural potential. Most of the avocado growing is done further south where frosts occur during the winter. These are not usually long stretches of sub-freezing weather—they are overnight blasts of cold that subside with the rising sun. Those lows, however, could damage avocados and ruin the crops—except for one clever trick that keeps the trees safe without heaters, smudge pots, greenhouses, or any of that fancy stuff.

The trees, rather than being planted on flat land, are planted on slopes.

Why does this make a difference? It's because overnight cold air flows downhill and settles in low spots. Instead of still, cold air settling around the trees and slowly freezing them to death, the icy atmosphere continually flows on downhill past the trunks. Avocados are not planted at the base of the hills in the valleys where the cold air is collecting—they are planted safely on the hill, and the cold literally washes over them without sticking around long enough to cause much if any damage.

Picture it as if you were pouring water from a pail. That flowing water will find the low spots and sink in there, rather than sitting at the high points.

Don't plant at the exact peak of a hill and don't plant at the bottom. In-between you'll find the sweet spot that means the difference between fruit and no fruit.

In flat areas the cold air settles and moves slowly around the landscapes, keeping us from using this method... but if you're lucky enough to have a hill, try the Chilean method of letting the cold air run past.

Swimming Pool Power

Now for another avocado story.

My friend Dave Taylor, owner of Taylor Gardens Nursery in Sparr, Florida, often had access to great, locally grown avocados in season. They weren't the cold-hardy variety, either. They were honest-to-goodness tropical avocados.

"Where in the world are you getting these avocados, Dave?"

He grinned. "A friend grows them right near here."

"How?" I asked, "those kind of avocados don't grow here. Greenhouse?"

"No," he answered, "swimming pool."

"Swimming pool?"

"Right," Dave said, "she has a house on a slope with three big avocado trees right behind it, then a swimming pool. They never freeze there—the pool keeps them warm."

I was impressed. Large trees, growing a good 100 miles north of their range, producing every year without a hitch. This method will work with just a pool, but my bet is that the combination of a slope, a pool and a warm wall combined to create a really sweet microclimate.

My pastor in Gainesville grew some very nice Cavendish bananas next to his swimming pool. Think of all that warm water on a cold night. The pool won't freeze here in Florida—and what you plant next to it is less likely to freeze as well. Think of it as a big, chlorinated space heater.

If you really want to go full crazy, build a greenhouse over your entire pool area like the Garden Pool people[5] did. They're raising all their food with a combination of hydroponics, chickens and tilapia in what used to be their swimming pool. That's turning a money-sink into a money-maker, though I don't think it would be any fun to swim in at this point.

Other Warm Spots

Big boulders (which are almost as uncommon as hills in Florida), patio areas, alleyways, bodies of water, masonry walls and other thermal-mass holding locations are good places to test out more tropical species. Some of the discoveries you'll make are surprising. I've planted trees in locations I thought would be warm, only to find that there were unfortunate currents of freezing air passing through on cold nights... and I've planted in areas I thought wouldn't be as good and had trees sail on through.

Here's another tale. I once helped a nursery customer plant a variety of fruit trees right on a beautiful lake. As we dug holes, the neighbor came out and watched over the fence for a few minutes. He was a somewhat portly guy with an intelligent face. In a calm voice he inquired, "what are you planting?"

"Well," I replied, "we planted some citrus over there along the hedge line where the oaks are, then over here we're planting some plums, pears, figs and peaches."

He shook his head. "Watch out, the cold off the lake is terrible. I lost all my citrus trees multiple times."

"Really? I would have thought the lake would be warm at night."

"Nope. The cold wind always comes from that direction," he pointed over the lake, "it gets funneled right through the hills, over the water, then blows up the hill here. Kills everything—gets really cold."

I had forgotten to take into account the fact that, though the lake was perhaps warm, it was also a wide open space that didn't block the cold winter winds.

The neighbor continued, "I did manage to grow some good citrus, though—want to see where?"

I nodded, intrigued. Stepping through the gate onto an immaculate lawn with tasteful landscaping, we walked together past his classic Victorian house and onto the front lawn. There he had a small orchard of productive citrus trees, including lemons, which are not known for their cold-hardiness.

"They all died behind the house, but in front of the house they do great."

Now I was seeing the big picture. The front of his house faced south; the back, north. His two-story home blocked the cold and created a sheltered garden that was likely a full growing zone warmer than the hillside leading down to the lake which was wide open to the dancing fingers of the north wind.

I went back and had a talk with the lady I was helping plant trees and we decided to put stone fruit and other cold-hardy species in the backyard and some of the citrus in her front yard. Other citrus we left beneath the oaks by the hedge which looked as if it would block the worst of the cold.

Takeaway point: pay attention to every aspect of your land... and be sure to talk to the neighbors who may know more than you do. Observation before installation!

Hard and fast rules of (green) thumb: don't plant tender plants out in flat, wide-open spaces. The middle of your lawn is a bad idea. Beside the shed is better. Beneath the edge of a large tree is also good. One thing I found when I built my North Florida food forest is that as the trees and shrubs became denser and the canopy grew, the frost damage decreased. I was gaining on the next growing zone thanks to the moderating effect of having all those plants together. I'll bet in another couple of years the new owner could tuck in some papaya out there and have them live through winter in decent shape.

The place where you plant your tender trees is probably the most important piece of the zone-pushing puzzle. After that, the care of the tree also makes a big difference.

Grow 'Em Big and Keep 'Em Happy

The cold-hardiness of a tree greatly increases as it matures. A little seedling may die right at freezing, yet an older tree might sail through a night ten degrees colder with minimal damage.

At this point, I must again plug Dr. David Francko's book *Palms Won't Grow Here and Other Myths*. It's worth purchasing just for his "Primer on North-by-South Gardening" near the beginning of the book. He shares the value of proper watering, fertilization, pruning practices, and other acts by the gardener that can make the difference between a happy tree in the spring or a rotting stump.

In relation to the age of a tree and its cold-hardiness, Francko writes:

> *Any gardener knows that mortality is highest in the first year or two after you install a plant. It takes a minimum of three or four growing seasons for a woody tree or shrub, for example, to develop a large and deep enough root system to sustain itself through droughts and cold winter temperatures. So, a first- or second-year palm tree is vastly less cold hardy than it will be after growing in the same location for several years. Cold-hardiness data in the literature and on plant tags are usually reported for mature, established plants only.*[6]

Get them big and healthy and they can take more stress. Makes sense, right? A healthy and active person may suffer a broken hip, heal up, and be back at the gym later that year, whereas a sicker person may never recover from the injury.

The cold of winter is a terrible thing for a tropical tree to face. It simply lacks the coping mechanisms and genetics to deal with freezing temperature. Unlike northern species which are designed to take the cold with such clever systems as anti-freeze compounds and light-cycle triggered periods of leaf drop and dormancy, your mango tree is naked to the cold. In order to survive and recover from each winter's lows, it must be in great shape heading into the cold and babied when it comes out of it. You can go ahead and prune out the dead wood once all danger of frost has passed, but—as I said earlier—don't succumb to the temptation to "make it look nice" while there is still a chance of frost.

Since bigger, healthier trees have a better chance of living, we need to give them optimal growing conditions when they're young so our task of keeping them alive through winter becomes less intensive. Just a few simple points of care will help your trees and shrubs to grow quickly. First, keep them watered. I know, that sounds like "duh" advice, but you'd be surprised how many people will pop a tree into the ground, water it a few times and then forget about it.

The power of watering and care shows itself when you compare two trees directly; one cared for, one not. I once bought a small jaboticaba tree (*Plinia cauliflora*) the same day someone I know bought one almost twice the size. Two years later, my tree was twice the size of theirs. Their tree had barely grown six inches after planting, despite the fact that they lived further south in a much better climate for this rare tropical fruit tree.

"Do you ever water that tree?" I asked.

"Well, it gets some rain," was the reply, "but the sprinklers don't really reach that area, so no, not really."

My own jaboticaba tree was in a large pot, the bottom of which was stuffed with rotten wood and compost. It got watered a few times a week and was fed with compost tea, chicken and rabbit manure, plus random applications of Epsom salts. I ended up selling it at a 300% profit when I moved... and it was six times the size it had been when originally purchased.

Watering makes a great big difference. Keep those trees free of drought stress when young—and old—and they'll be able to get through winter better than if they've been scrabbling along. "What doesn't kill you makes you stronger" does not apply to trees.

Now before you think I'm some sort of super gardener because of my perfect jaboticaba I must tell another story.

The same aunt I used to have who loved pruning also was great at getting trees to grow quickly. We both bought peach trees at the same plant show and planted them around the same time.

My tree barely grew in its first year whereas hers doubled in size. I have never seen a peach tree grow as quickly as hers did. It went from a 1-inch diameter trunk to a 4-inch diameter trunk within two years.

I asked her how in the world she'd gotten that tree to grow

so quickly. The answer was "lots of cow manure, keeping all the grass far from the trunk, and running the hose on it regularly."

She would run that hose, too. Every day she had it running on a trickle at the base of one of her fruit trees. They never dried out, never had to fight with grass roots and never wanted for any fertilizer.

That's why her peach tree way outpaced mine. I watered perhaps once a week, didn't feed it much, plus the grass and weeds got away from me. Forgive me—this was six long years ago when I was young and foolish.

Keeping the grass back is a big factor many gardeners never consider. When a gardener lets the grass grow right up to the base of his fruit trees, he's starving his tree of water and nutrition—even though it may "look nice" that way. It's been estimated that grass roots grow twice as fast as tree roots and some species of grass are also allelopathic, meaning that they exude substances that can inhibit the growth of other plants. Clean up the weeds and grass, then mulch to keep them from coming back. I like to keep a clear area at a 4-foot radius from the trunk of my newly-planted fruit trees. A lot of the feeding is done in the first couple of inches of soil which is also where most of the grass roots will be. Don't let them fight—get rid of grass, and your trees will grow much faster.

Feeding is another factor many overlook. Trees don't complain all that much if they aren't fed. They won't whine like a dog or cry like a child. They'll just sit there and slowly starve.

Most of a tree's "food" actually comes from the sun through photosynthesis, yet without certain nutrients the many complicated systems inside the organism start to break down. The big three nutrients are nitrogen, phosphorus, and potassium (NPK), but there are also lots of micronutrients that are needed as well. A "balanced" fertilizer may have three equal numbers on the bag, but it's not ideally balanced for any individual plant. Proper tree fertilization is beyond the scope of this little book, but just think of a tree's nutrition in terms of how you'd raise a healthy child. A broad range of inputs is better than just pure protein, carbohydrates and fats. I give my trees compost, compost teas (I have a lot on composting for maximum soil fertility in my book *Compost Everything: The Good Guide to Extreme Composting*), manure from our own farm (*not* manure from other farms, as it's often contaminated with nasty long-term herbicides that will kill your trees—I also cover this problem in *Compost Everything*), chopped-and-dropped weeds, grass clippings, and twigs as mulch, coffee grounds, eggshells, bone meal, blood meal, Epsom Salts, fish emulsion, elemental sulfur, micronutrient fertilizer blends, and other bits and pieces. Even just dropping chunks of logs on the ground in rings around your trees will foster the growth of good fungi that help your trees feed more efficiently.

What I don't do is feed the trees late in the year. New growth near winter is a bad idea, and fertilizing past midsummer will often get your trees to put out a flush of tender growth that frost destroys much easier than older growth. Even if your trees look hungry, just keep them well-watered and wait. If they really look starved, I might give them some magnesium in the form of Epsom Salts or even some micronutrients, but I'd stay far from the nitrogen! Nitrogen in the fall is like espresso before bed. Bad idea.

Get your trees to grow quickly in season then keep them well-fed and your prospects for winter survival will improve.

Don't Forget to Cover

Finally, don't forget that those little trees are much more likely to be killed on a cold night than a larger tree would be. If your plant label says "hardy to 25 degrees", don't trust those numbers on a newly planted specimen. 32 degrees may very well kill a young tree.

Put barrels of water next to the most tender specimens and cover well. If the weather is projected to get "close" to freezing, cover just in case. It would be a shame to lose a beautiful young tree with so much future potential just because it was too much trouble to go outside and toss a blanket on it. For the first few winters at our North Florida home, we were tied to the homestead through the winter because of our many tender young citrus trees. Though we were often invited to visit with family further south, sometimes we had to say no because of approaching cold. This paid off, as many of those trees are now large and covered with abundant, valuable fruit—and they're much more hardy than they were as little 4-foot trees. You can often afford to leave a big tree alone through a frost but you can't assume the same with a young tree.

A little work now leads to sweet rewards in the future. Give your trees the best chances of survival by taking care of them when they're small, and one day you'll be standing beneath the shade of a tree that can support itself.

Chapter 9

Your Secret Slice of the Tropics

Now we get to one of my favorite discoveries in my zone-pushing journey: the power of a south wall. Little did I know when I moved in that my house in North Florida had about 70 square feet of Miami climate just outside the back door.

Growing tropical plants outside of the tropics is usually accomplished with sunny windows or greenhouses. Yet when you have a south-facing wall it's more than possible to grow plants you never thought could survive in your region.

I came to this discovery over a period of time as bits and pieces of information presented themselves and were filed away until they suddenly came together in one epiphany. Though I didn't yell "*EUREKA!*" I could have.

I did plant a Eureka lemon beside the wall, so that has to count for something.

The story starts back when I was in Tennessee. I planted a somewhat rare fig in the corner of my backyard in the "L" created by the south-facing wall of my house and a large shed

I had installed as my painting studio. I'm sure I had read something about south-facing walls being good for tender plants but I have no idea where I came across the information originally. I just figured "why not?"

I'm glad I tried it: one night the overnight low hit 14 degrees and that little fig sailed through without freezing.

I was intrigued. *Wow,* I thought, *I was hoping it would do well in that protected location... but I had no idea it would do that well!*

When I sold the house after our stint in Tennessee and moved back to Florida, I figured our yard would grow most everything I wanted to grow... until the first winter took my mangos and papayas and taught me differently.

While I was visiting a friend who also lived in North Florida, he showed me a key lime tree he was trying to grow. It was pretty sad. The tree was nothing more than a few shoots returning from the ground. "It freezes down every year but it always comes back," he said, "I don't think we're going to get any limes any time soon."

If you're familiar with the cold-hardiness of various citrus, you know there's a spectrum of what they can handle. Kumquats can take a lot of cold... and key limes can take almost none.

Suddenly growing key limes in North Florida became painfully important. It needed to happen.

Serendipitously, I was given a key lime tree when someone I knew lost their home to a foreclosure. Good for me, though not for them.

I decided to plant it next to the south-facing wall on the back of my house. I had a little plaza there I had created from brick pavers, so I pulled up a little square of pavers about three feet from the wall and planted the 4-foot tall key lime tree.

Then winter came, and with it at least one night in the 20s. Within a few days of the frost, it became quite apparent where the south wall microclimate ended: almost directly in the center of the tree!

The half of the tree furthest from the wall was all scorched brown by frost, the half closest was green and untouched.

It was such a perfect line of damage it looked like someone had spray-painted half the tree brown. In retrospect, I really wish I had taken a photo. Or painted its portrait. The shepherd's needles and other weeds in the area mirrored the damage. Within three feet of the wall they were unscathed... farther than that, they had melted in the cold.

Three feet was too far from the wall for that poor key lime, I decided, so I planted it at just six inches from the life-preserving cinderblocks of my newly mapped microclimate. It took some serious pruning of the back half of the tree to make it fit but I pulled it off. To keep the branches from spreading outwards too much, I made a simple frame of electrical conduit and bolted that to the wall, then tied any errant branches back as they grew.

I never had to protect that tree again. Winter after winter it sailed through the cold, blooming in cold weather and providing us with enough limes for pies every Thanksgiving, not to mention Margaritas and lime slices for the occasional bottle of Corona. There have been 12 degree nights when the tree sailed through with only a touch of frost damage on the few branches and leaves that stuck outside the protection of the wall. I keep pruning and tying the tree back to ensure it stays growing happily. The method is called "espalier," which is French for "a total pain-in-the-neck way to grow trees," but it satisfies your inner control freak. It works!

I now knew where the boundary between temperate and tropical lay. It was time to conduct further experiments.

The first addition other than the key lime I made to the south wall was the guava that had frozen to the ground and come back. It needed a better home. When I saw the shoots returning in the spring, I carefully transplanted it about eight feet down the wall from the key lime where it could finally grow in peace.

I also had a couple of lemon trees in pots that needed a home. When I started my nursery, I discovered the regulations on citrus meant I couldn't keep any citrus trees in pots without a special license. These two lemons had been going in and out of the greenhouse every winter and bearing us lemons. Since that was no longer possible, I planted them on either side of the key lime and pruned as needed.

Another addition was some papaya trees, planted tightly to the wall where they could bear their musky melons without being melted by the cold. A few pineapple plants and a bunch of banana trees rounded out the far corner of the wall. The bananas had the additional benefit of being watered by all the run-off from our kitchen sink, helpfully piped to their bases by a rigged-up piece of PVC with lots of holes drilled in the last few feet of its length.

Later I added a Kaffir lime and another guava, both of which did well.

After a couple of years of seeing how well these plants lived through the cold, I got even more ambitious.

I planted a coffee seedling (*Coffea arabica*), some longevity spinach (*Gynura procumbens*), Okinawa spinach (*Gynura bicolor*), Cuban oregano (*Plectranthus amboinicus)*, katuk (*Sauropus androgynus*), eddoes (*Colocasia antiquorum)*, Malabar spinach (*Basella alba*), waterleaf (*Talinum fruticosum*), and then, just for the heck of it, a black pepper vine *(Piper nigrum)* at the base of the original key lime tree.

I dubbed the space my "Miami garden," and it outperformed my wildest expectations.

All of the plants lived through the following winter—even the coffee and the black pepper. I found that startling as both of those would die in winter even a couple hundred miles south.

It seems like magic, but it isn't. A wall—particularly a stone, brick, or cinderblock wall—soaks up heat during the day and slowly radiates it back out again at night. Rocks, ponds, and the ocean do the same thing. Again, that's why you'll see plants growing near the shore that would never survive the cold even a few miles farther away. Note: the overhang of your roof is a big help, too. The larger the overhang, the more protection you get from the chill of space above. That shade, like the shade of a tree, makes a difference. That said, you don't want a lot of shade over your south-facing wall. Large trees will keep the sun off the wall and lower its power to protect your tropical plants, so you may want to thin out or remove any limbs keeping the limited winter sun off your tropical garden strip.

I've been amazed by how well a south-facing wall works as a tropical garden, to the point where I've wondered if you could construct walls like this on purpose rather than just taking advantage of a happy accident of your home's construction.

Proving there's nothing new under the sun, it turns out the French already had the idea centuries ago and employed it to great effect:

> *Established during the seventeenth century, in a prominent neighbourhood of the eastern edge of Paris known as Montreuil, a 300 hectare maze walls and agricultural plots provided a unique and unlikely microclimate for [peaches], normally suited for cultivation in warmer areas such as France's Mediterranean coast.*
>
> *The peculiar architecture, known as "Murs à pêches;"* wall for peaches, *served to protect peach trees planted near the walls and adapt them to a much colder environment than the fruit is typically used to. The 3 meter high walls were more than half a meter thick and coated in locally sourced limestone plaster, giving them a high thermal inertia and the ability to store heat. Their intentional north-south orientation allowed solar energy to be stored*

in the walls during the day and transmitted to the trees during the night, preventing them from freezing and accelerating the ripening process. Within these walled orchards, temperatures were typically 8 to 12 °C higher than outside. With the legendary Parisian marketplace, Les Halles, nearby, the peach farmers of Montreuil had a guaranteed market for their product. The Parisian "pêchers" of Montreuil became famous within high society, supplying peaches for the court of Versailles and French nobility in the 17th century. Even the Queen of England, the Prince of Wales and Russian Tsars came to the peach orchards of Montreuil to taste the unique varieties of Parisian peaches."[7]

So why haven't most of us gardeners heard about this method yet?

You have now!

Imagine growing starfruit, dwarf mangos, jaboticaba or acerola cherries outside without protection… in the middle of Texas. Or growing Japanese persimmons against a wall in New York. Or getting loquat fruit in Oregon. Chances are, you can with a south-facing fall and some careful pruning. Thank God for that key lime tree lighting the way!

By the way, if you live a little farther north and are worried you still won't be able to pull off a key lime tree, there is another option: the key limequat.

Key limequats are a pretty little citrus fruit with good flavor, originating from a cross between key limes and kumquats.

The trees are short and moderately cold-hardy, unlike the very cold sensitive key lime.

A few years before I quit recommending citrus, I planted a key limequat in my yard in a semi-shady area. Last year it gave me a nice hat-full of fruit—plenty for pies and drinks.

Key limequat rinds are sweet while their centers are quite sour. After you eat one, it's very much like you took a bite of sour key lime pie.

The taste isn't exactly the same as a key lime, as key limequats tend more towards the sour without that tangy bitter undertone that gives key limes their unique flavor profile.

Key limequats are a non-demanding citrus that require little more than a small corner of your yard and some water to get established.

As covered before, I wrapped my tree on freezing nights with thrift store sheets and blankets. The key limequat always came through fine, unlike some of my oranges which usually suffer some frost damage despite the blankets.

Hoe away weeds around the base of your key limequat and it will reward you with faster growth. I mulched mine even though that's not the recommended practice for citrus.

Key limequat trees can take some shade, can grow in pots and they'll also do great planted up against the magical south wall of your house if you live in a colder area.

The biggest problem you'll face in growing key limequats in many areas is the greening virus. Greening kills all citrus and it's spread throughout my home state of Florida. If you're in a location where citrus greening isn't a problem, great.

Otherwise, the life of any citrus tree you plant is likely to be cut short. It's basically incurable, though some strides have been made in recent years with heat treatments and nutritional supplementation.

Doesn't that bring a tear to your eye?

I hate what citrus greening has done to the citrus across my home state and to the farmers that rely on citrus for their livelihoods. I generally blame monoculture farming for this disease, yet I understand why citrus is grown the way it is for the sake of management and harvesting.

So—should you plant a key limequat in the face of greening? That's up to you. It may live a long productive life... or it may need to be pulled out in four years. I have read that kumquats are more resistant to greening than some citrus, as are key limes. This may simply be because they're less attractive to the psyllid that spreads greening, rather than any intrinsic resistance to the disease itself.

Thus far, my Florida citrus trees have done fine. We'll have to wait and see.

But... back to the south wall: just do it. If you have azaleas there, tear them out and plant something delicious, edible, and exotic instead. That wall is a serious resource that will surprise you with its power to stave off the chill.

Give it a try. You'll be amazed.

Chapter 10

Breeding for Cold Tolerance

NOTE: Now we're past the area of my own zone-pushing experience and we're getting into crazy territory.

I will state right at the top of this chapter: I'm not a serious plant breeder, I'm a backyard plant breeder. I'm not an accredited scientist, and I lack a degree in anything related to horticulture. However, I am really good at pattern recognition, and I've been gardening for thirty years. I'm also a dreamer, and I enjoy inviting you to share my dreams.

So... why not dream the impossible dream of breeding plants for cold tolerance?

Let me tell you a story. A strange tale with an unsatisfying ending... but a tale, that if true as told to me, is wondrous.

I was standing among the flowers and vegetables in my little cottage garden in Tennessee when I saw him. A middle-aged, smiling gentleman with a small handful of flyers of some sort.

"How can I help you?" I asked.

"I'm just out sharing the good news with some of my neighbors. I have to ask you, if you were to die right now do you know where you'd spend eternity?"

"Yes," I replied, then assured him I relied solely on the sacrifice and imputed righteousness of Jesus Christ to secure my passage from this life to the next.

"That's wonderful!" he said, then handed me a flyer for his church, a Methodist congregation nearby. "I was working on my garden this morning, and I suddenly felt called to go work in the Lord's garden." He looked around my gardens, took in the wild profusion and continued, "by the way, you have a great garden of your own here!"

I thanked him and we got chit-chatting about gardening. At some point I revealed that I was from Florida and really missed being able to grow citrus.

"Well, I know a guy that's growing citrus right near here," he said, "he's got lemons in his backyard."

I was stunned. How was that even possible?

"How in the world? Does he have a greenhouse or something?"

"No, he just kept planting seeds, over and over again for years until he got a tree that lived, then he's been starting more off that one."

We were near the border of zones 6 and 7. This was preposterous. The only citrus I knew that would survive that far north was the barely edible trifoliate orange. And yet, here's a door-to-door missionary sharing some good news I really didn't expect to hear.

Then he dropped the bomb.

"I can get you one."

This was what every plant geek wants to hear when they've just discovered the craziest, most awesome plant ever.

"I would love that!"

He left to continue canvassing the neighborhood, and I kept working on my garden, wondering about this lemon story and if I'd see the guy again.

A couple of days later he returned with a potted lemon seedling that was perhaps two feet tall. He'd also brought some herb plants with him.

I was amazed and thanked him profusely.

Later I planted the little lemon near the fig that had survived winter... and then the Nashville flood came.

With the flood came a huge amount of water through our backyard, tearing out all of my garden beds, mulch, the spring gardens and years' worth of precious topsoil accumulation. It even bent one of my apple trees over to a 45 degree angle.

As I watched the muddy river tearing past my back steps, I cursed the rain. The yard had flooded before and wrecked our gardens... but this was the last straw. I was done. The Town of Smyrna had rebuilt the drainage ditch in the backyard the previous year... and then dumped more water into it off a nearby highway. It couldn't handle the water, and I couldn't stand losing my garden again. Plus, the town was hell-bent on controlling the lives of its citizens and keeping us from backyard chickens or even allowing some wildflowers to grow. They threatened to fine me for the height of the tulips I'd planted in the front lawn, claiming they were "undergrowth over eight inches." To heck with them all. I was done. They could keep their stingy, dingy town.

So we moved... and I forgot all about the lemon in the mad rush of repainting the house, real estate showings, the closing and the big move back down to the countryside of my home state of Florida where I could grow things as tall as I wanted and keep goats in the front yard without incurring the wrath of petty bureaucrats.

The next year I was talking with someone about cold-hardiness and I remembered that tree that had been so precious... and that I had stupidly left behind.

I still kick myself. I visited Tennessee the next year and drove past the house but couldn't tell if the little lemon was still there. Sneaking into the backyard seemed like a bad idea and there was no one home to ask. Rats!

I also wish I knew who that missionary was and who he got the seedling from. Somewhere in Smyrna there allegedly grows at least one cold-hardy lemon. At least that's one good thing in that miserable town.

So, what can we learn from this strange tale? First, don't move to Smyrna. Second, when a genetic sport comes your way, *hold onto it!*

If that lemon tree really was that cold-hardy and capable of fruiting in Tennessee, it was worth an absolute fortune. We're talking millions if it were patented. I should have tracked down the original owner and let him know. And I certainly shouldn't have moved and left my seedling.

However, it may be that the tree was not all that cold-hardy and had just been lucky for a few winters or planted in a unique microclimate. It also may have been a lousy fruit, like a trifoliate orange. Or the door-to-door missionary got the story wrong.

I have no idea. But I do know that there are genetic codes hidden in plants like treasures waiting to be unlocked. If one of those treasures is cold-hardiness in a particular species, you aren't going to find it without planting lots of seeds. The zillions (this is a technical term used by scientists) of genetic combinations possible will only exhibit themselves if people do the time-consuming work of growing trees from seed.

Our apple varieties in all their variety originally came from a seedling tree or a chance bud mutation. Broccoli, cabbage, kohlrabi... someone selected their traits over time.

If there are cold-hardy genes hiding in mangoes, for instance, they will come forward in a seedling. You can plant lots of seeds. And also, always be looking for trees growing out of their normal range. I once found an absolutely monstrous Guanacaste tree (*Enterolobium contortisiliquum*) growing by a North/Central Florida roadside far from its original home in the tropical rainforest. I saved seeds and planted them in my food forest as a nitrogen-fixer where they sailed through the winter.

If the genes for a particular trait you seek do exist, they may already be in a tree growing in a backyard, an empty lot, or the woods. Keep your eyes open, and ask lots of questions.

Join local plant groups, such as the many found on

Meetup.com, and ask questions like "Does anyone know of a mango tree growing around here?" Someone may. See if you can go and see it, then try to ascertain if that specimen is in a microclimate... or if you've found a frost-resistant pot of genetic gold at the end of a tropical rainbow.

    You never know. Stranger things have happened.

Chapter 11

Strange and Marvelous Crops to Try

Now that you have some wild and wondrous frost-protection methods in your gardening toolkit, it's time to dream about what you can try growing in your area. When I was in Tennessee I was secretly rooting for global warming to be true just so I wouldn't have to work so hard to grow warm-climate trees and shrubs. Yet when you can make your own climate in your yard the options that open are quite exciting and don't require the submersion of Miami condos by rising sea levels. Personally, I believe we're more likely to face a new ice age than a hotter earth... but whatever you believe about climate change, you have more control over your options than most gardeners realize. Adding just a few degrees can be life-changing.

I've shared many of the trees and shrubs I managed to pull off far from their "natural" range so far. Here are some more details on various tropical edibles worth trying out. Some are a big stretch, some are pretty simple.

Avocados

There are avocados growing in Gainesville and frost-prone areas of Texas and Mexico. Good avocados, too. Though varieties like Hass can only take cold down to the mid 20s even at maturity, there are avocados now on the market that can handle weather in the low teens without being killed. I purchased multiple cold-hardy seedlings at the Gainesville farmer's market that were second generation trees planted from the pits of a mature and productive tree. Even outside the city where it's colder you can still pull off good avocados.

Someone once asked me if their tree was a cold-hardy type. I plucked a leaf, crushed it, inhaled deeply and responded, "Yes, this is a cold-hardy type."

I think they thought I was practicing some sort of horticultural sorcery. How did I know it was a cold-hardy type? The Mexican cold-hardy genetics in avocado trees can be determined in a rather strange way. Pluck a leaf and crush it. If it smells like anise, you have cold-hardy genetics in that tree. If it doesn't smell like much of anything, it's a typical tropical avocado. Somehow the genes for anise scent and cold-hardiness are linked. It's a strange world.

Types to consider if you live in zone 8/9 include Lila, Brogdon, Pancho, Fantastic, Joey, and Day. Lila is tolerant of cold down to 15 degrees... and maybe even lower. This means you can probably pull it off in big chunks of Texas, Mississippi, Louisiana, Alabama, Arizona, and other places not generally known for their avocado production.

The biggest worry for me with avocados is their susceptibility to Laurel Wilt Disease. I've seen some great avocado trees destroyed by this disease, and if it keeps spreading, avocados in both warm and cool areas will be much harder to grow than ever before.

Bananas

Growing bananas and keeping them alive through a winter with repeated frosts isn't rocket science. It's not even hard. It doesn't require sheets, Christmas lights or a greenhouse.

All you need is a little patience, a somewhat sheltered location, lots of nutrition and water... and for you to put away your machete... and you'll be able to grow bananas a good couple of zones north of their proper range, no matter what everyone says.

Now don't think you're going to plant a little banana pup and get fruit the first year. Bananas quit growing when it gets cool out. We're talking at 70 or below... they just kind of hang out and wait for it to get hot again.

If you live in a place that has cool winters, during those few months your bananas are just going to sit. Frost damage will also set them back, which leads me to my next point.

Winds and cold nights will wreak havoc on your trees. If you have a south-facing wall with some sun, plant them there. I love south walls and you will too, once you see what you can pull off.

Other locations that are good for bananas are right beside a swimming pool, next to a fence that blocks the wind, or under the outer edge of a large shade tree so they get some sunlight but also some canopy protection from cold nights.

Shorter bananas are easier to shelter than taller varieties, but I still grew 16-foot Orinoco bananas at my place, and they do great despite some nights that have reached down into the teens.

Bananas like lots of fertilizer, so feeding them up like crazy during their growing season will get the trees closer to fruiting. Banana fertilizer can be compost, urine, chemical fertilizer from a bag, or whatever kitchen scraps wash down the sink. Septic tanks are also beloved by banana trees. Just know this: they're hungry trees.

Bananas are also very thirsty. I read somewhere that they like at least 100 inches of rain a year. Plant them at the base of a gutter and/or make sure they're getting water regularly or they stall out. Graywater systems are great for bananas.

Banana circles are a great way to combine both water and fertilizer in one. This gardening method is covered in *Compost Everything* but I'll just give you a quick overview here.

A banana circle is a permaculture idea where you dig a pit with a berm around it, throw a bunch of highly nitrogenous materials in the middle, then plant bananas in a ring around it on the berm. The roots run into the middle and you continually dump in new materials as they become available. Kitchen scraps, urine, manure, ramen noodles... whatever. The bananas will eat it. I ran the kitchen sink pipe directly out into my clump of bananas and they loved it. All the scraps that went down the drain were eaten by the bananas.

Another thing: don't cut back your bananas!

For some reason, folks where I lived in North Florida like to chop back banana trees after a frost, along with removing all the dead leaves. Don't do that! First of all, only remove the dead banana leaves after all danger of frost has passed. Secondly, cutting the trunk back will set the tree back a season or so and keep it from fruiting as soon. I know, the tree looks ugly—let it look ugly for a while. Usually, new shoots will emerge from the center of the tree once the weather warms up—and often, a bloom and bananas will follow. If you chop them back this usually won't happen until later in the year and you may lose your bananas to frost.

Bananas are non-seasonal trees (actually, they're not even trees... but I digress), meaning they fruit when they feel like it. This could mean that they decide to put out a bloom in September and then as it cools off in October, you have a little hand of bananas hanging there and not maturing because the weather has gotten cool... and then in November or December, a freeze will take out the fruits before they have any kind of chance to ripen.

Ideally, your tree will outgrow the frosts of winter and put out blooms in the spring and early summer, ripening up by fall and frost season.

Gardeners regularly ask me how and when to harvest bananas. Harvest all of a stalk of bananas by cutting down the complete stalk—but only do this when the fruit have all filled out and plumped up and the top few are starting to turn yellow or at least green-yellow. It's not quite a science, but it's close. Catch them at the right time and bring them indoors or hang them on a porch. Once you harvest your stalk of bananas, cut down the entire "tree" that bore the fruit—it won't bear again. The biggest pup beside it will replace it.

If you divide off banana pups as the clump grows, you can plant them in different areas of your homestead to see how far you can push them—and it won't cost you a dime. Long ago, farmers and hobbyists bred the seeds right out of most good eating banana varieties. This is a good thing since banana seeds are like birdshot and aren't kind to your teeth. What this means, though, is that bananas can only be cultivated by clonal reproduction. Just divide a little banana pup off the base of a clump of bananas carefully with a shovel, getting a chunk of the main root system while NOT cutting through the little banana's trunk and leaving the roots behind... and you're set. Bananas grow lots of little pups if they're happy and one banana can easily turn into a clump under the right conditions.

We've had good success with both Orinoco and Raja Puri bananas. Dwarf Cavendish will also produce, though it's a bit slow and prone to frost damage. Other varieties are worth testing, however, since most of the work with bananas has been done in the tropics. My friend Joe Pierce co-owns the Mosswood Farm Store in Micanopy, Florida, and has a great edible plant nursery there, along with a bunch of banana varieties he's testing. I asked him for his opinion on varieties. He writes:

Mysore, Blue Java/Ice Cream Banana, Latundan/Apple, Dwarf Orinoco, Dwarf Namwa, have all produced for me after going through frost. Thai black (though it survives) is a seedy ornamental and the Red Dwarf I got from you did not come back after frost, pups survived mother did not. Praying Hands is hardy also but more of a novelty than what I would call good eating. There are many more that I've seen at Oliver Moore's fruit after being frost bit but I haven't personally harvested them in my trials. Such as Pisang Ceylon/Improved Mysore, fhia25/plantain hybrid, High Noon, Mona Lisa and Gold Finger. My favorite for flavor and ease of growing is the Dwarf Namwa which sets large amounts of fruit on a short stocky plant that does not require staking in order to support its heavy bunches. It only gets about nine feet tall which makes harvesting a breeze not to mention the fruiting bunches are right at eye level and the flowers are intoxicatingly fragrant. (Fruiting) is just a matter of how old the mother plant is, fertilizer and water. Supposedly the Raja Puri fruits at the youngest stage which is why it's commonly grown. Veinte Cohol is another young fruiting variety that I'm working with but have not had fruit yet.

I can vouch for the rapidity of Raja Puri's production. A friend planted some two-feet tall plants by her house in the spring and took good care of them. They all grew to eight feet and one even fruited that same fall.

I knew a fellow in Tennessee who grew a banana tree in his front yard. Every year he'd dig that tree up in the fall, carefully wrap up the plant in burlap, then lay it in the earth beneath his house until spring; when he'd plant it again. It was a lot of work, sure; but it was insanely cool that he had a banana tree growing way out of its range.

If you live too far north to zone push fruiting bananas successfully but you still want some bananas, *Musa basjoo* is a good bet. It will live even in very cold climates if you mulch well, though you won't get any food from it. It still looks tropical, and that counts for something.

Dwarf varieties in pots should be able to fruit in cold climates if you drag them into a sheltered location for the winter. Maybe big pots with casters on the bottoms? That would be awesome.

Black Pepper

It took me quite a while to track down black pepper plants (*Piper nigrum*) I could test in my North Florida gardens, but I eventually found some little plants on eBay. Over several years I grew those little plants up into larger, sprawling potted specimens that I kept in my greenhouse, yet they never fruited for me. After I developed the south wall garden and saw its success, I decided to plant a little black pepper vine at the base of my key lime tree to see if it would live through the winter. From what I'd heard, black pepper was not a cold-hardy plant by any stretch of the imagination. It's a true tropical, native to the hot and humid tropics of India. However, I was cultivating its tasty cousin the "root beer plant" (*Piper auritum*) in my front yard food forest, and it was happily growing and blooming each year as a dieback perennial. Why not try the pepper? My black pepper on the south wall lived through the winter and is now over six feet of vine, branching and twisting here and there around the key lime tree and into the lemon to its right. Still no fruit, though. Maybe the new owner will get some. The big surprise is that it's not dead.

Cacao

Growing chocolate in your backyard? Ah, what a glorious dream that is to many temperate gardeners. Sadly, unless you're a super gardener with a good greenhouse, it's going to remain a dream. Cacao has zero cold tolerance and can die before you even near freezing temperatures. It's also very picky when it comes to humidity. A former cousin of mine tried growing a cacao tree indoors in a warm, well-lit room, and it dropped all its leaves in several weeks. He simply couldn't keep the air humid enough. This additional constraint means that even the cacao being grown in Homestead's marvelous

"Fruit and Spice Park" is kept in a moist greenhouse. I wish I could tell you something different, but this plant is a no-go. You might try carob, however, for a much hardier alternative to chocolate. Allegedly it grows in the same range oranges do, which is a much larger zone than cacao can handle. Carob also needs much less water to thrive.

Cassava

Cassava has a range that reaches farther than commonly recognized. You can grow it right up to the warmer portions of USDA Zone 8, though getting a good harvest takes two years rather than less than one as found in the tropics. Where I live in Central America you can pull roots in six months or less. In North Florida it usually took me around eighteen months. Why? Because cassava roots survive beneath the soil through winter but the entire top of the plant freezes to the ground. This loss of leaves and stems requires the plant to grow back in spring to keep feeding the roots before they enlarge to a decent size for the cook pot. I have been told by other Florida gardeners that they pulled off good roots just within one year. Maybe they used more compost than me. Or actually watered their cassava patch.

The main difficulty with growing cassava is finding the cuttings to plant in the spring. If you're in an area with frost, you won't have any viable pieces of stem to chop up and plant when you need them in the spring. I figured out that you could simply cut down some cassava canes in the fall before the first frost and store them in a pit covered with straw and a tarp through the winter, then pull out the still-living cuttings in the spring, chop them into sixteen-inch or so sections and plant them out. My friend Rick once managed to keep cuttings alive in a trash bag in his garage all the way through the winter.

If you plant cuttings in the spring and leave the plants in the ground through the winter, you'll have good harvestable plants the next fall. It's not ideal—but hey, fried homegrown *yuca* is pretty darn rare north of the Rio Grande!

Chaya

Chaya, also known as "Mexican tree spinach," is a relative of cassava which is instead grown for its nutritious and prolific greens. They are delicious though uncommon in most gardens. The leaves need to be boiled before eating or else they're poisonous, which also limits its appeal to those wussy gardeners who are afraid of death. Just cook them. It's really not hard.

Chaya can get huge in zone 10 but is a dieback perennial in zones 8 and 9, where it rarely gets taller than six to eight feet. The first year is important to its survival as the roots aren't as developed. If I can get chaya to live through its first winter it does better in the subsequent ones. It seems a bit more susceptible to death by winter than cassava but can also produce leaves in shade whereas cassava won't make anything worth harvesting in the shade. And remember: shade is frost protection! Pop some chaya into sheltered locations, and you'll enjoy hearty greens through all the warm months. If you live farther north, you'll be stuck growing it in pots.

Cherry of the Rio Grande

Is it too warm for you to grow Bing cherries (*Prunus avium*) and too cold to grow Barbados cherries (*Malpighia emarginata*)? You may be in luck: the "cherry of the Rio Grande" (*Eugenia involucrata*) is the in-between cherry. Unfortunately, it isn't super cold hardy and only takes you a couple hundred miles north of Barbados cherry territory without protection... but you know now how you can stretch that at least another hundred or more, right? Push the zone!

Though it's not really a true cherry (granted, neither is the Barbados cherry), I was surprised by how rich and cherry-like the flavor was when I tasted the first fruit borne by one of the shrubs I planted.

Citrus

Citrus has a pretty big range when you think about it. Varieties will grow all the way from the equator and up into the middle of the United States. The very best varieties are tropical or close to it, though. The most cold-hardy citrus is the trifoliate orange which has a ragingly bitter fruit you won't enjoy eating fresh, though it can be made into marmalade if you have plenty of sugar on hand. Like, buckets worth. This citrus will live through a zone 6 winter, which blows my mind and makes me wonder if some hybridization could be done that would transfer this trait to a more edible variety. Other somewhat cold-hardy varieties include the Satsuma, kumquats, calamondins and the orangequat, though none of those really do well any colder than zone 8. I'll bet it's possible to grow some edible citrus in zone 7 if you use the tricks in this book, though. Try a south wall and keep the tree covered with a barrel of water next to it during frosty nights for the first few winters until the tree gets bigger. If all else fails, citrus makes a fine container plant and can be planted in a pot and brought in through the cold.

Coffee

Along with chocolate, coffee is another dream plant for gardeners seeking the exotic. Unlike cacao it's not too hard to grow. I kept a coffee tree growing in a pot for years and harvested the fruit every fall. After a few years of keeping it in a greenhouse, I decided to try my luck planting one against the south wall of my house. To my great surprise, that little seedling survived and grew. I'm not sure if it will survive multiple winters there but I have a gut feeling it will. The tree is almost flat against the warm wall and showed almost no frost damage despite temps down to 20ºF one night. This is remarkable for a truly tropical plant like coffee. You can read more about growing coffee (and tea) in my booklet *The Survival Gardener's Guide to Growing Your Own Caffeine* if you'd like to dive down the caffeinated rabbit hole. If you live up north, treat coffee as a houseplant. It will fruit indoors, unlike many trees, making it worth the experimentation.

Coconuts

Give me a beach and some coconut palms, and I'll be happy. Heck, I'm happy with either by themselves. Coconut palms are so important I cover them in detail in Appendix B. Go read my thoughts on the possibility for coconut zone-pushing there. Then come back to this section and we'll get on to Cuban oregano, which is rather less exciting but okay for something that isn't a coconut.

Cuban Oregano

Cuban oregano (*Plectranthus amboinicus*), also known as Mexican thyme, big thyme, Mexican mint and various common names, is a fleshy, fast-growing large-leafed herb with a flavor similar to regular old boring oregano. It's quite zippy raw, but when sautéed or added to soups it's mild enough to be used as a flavorful green.

Cuban oregano starts quite easily from cuttings. All you need to do is break off a chunk of the stem and stick it into the ground. A few weeks later, it will put on new growth. I've grown them in the ground by my south wall, in the greenhouse and in pots. A friend of mine grew some on her kitchen windowsill. They like some sun but can handle the shade as well. What they can't handle is the cold. If you wish to grow this useful herb, you need to push the zone or root some cuttings in the fall and bring them in through the winter. If you keep the plant from being killed by cold, it will reward you for years with its abundant culinary bounty.

Ginger and Turmeric

Most people living outside the tropics have no idea how ginger grows—or that turmeric is a type of ginger. It's actually quite easy to cultivate, even if you live up north, provided you don't let it get too cold. Grow ginger and turmeric in the shade as they do not like the sun. This is good for zone-pushing, as the shade of a tree gives the plants even more protection. Shade-loving is a good thing when you're growing tropicals outside their range!

In North Florida I planted roots any time during the year. They will grow through the warm months, then the plant starts to die back in fall and go dormant. Whatever is left above the ground freezes down, then new shoots emerge from the ground when it's nice and warm. Turmeric really takes its sweet time coming back in the spring, often waiting until June to make its appearance. For the first few years I grew turmeric I thought it had died in the winter... and I was always wrong. The roots grow about four to six inches beneath the surface of the soil so the cold never reaches far enough down to kill the plant.

I usually plant ginger in spring, let them die back the first year, let them grow again all through the following summer, then dig some roots in the fall. I'll leave a piece of root in the ground so they come back again the next year. If you take extra good care of ginger you can harvest it in the first year but I never paid much attention to it and just took pieces from my food forest as needed once they'd been in the ground for two years.

I once helped my friend Cathy dig up a big two-year-old patch of ginger, and we were amazed by how many roots we pulled from the ground. It can grow into zone 8 and probably zone 7 if you mulch it well. If the ground freezes solid, the ginger will surely die, so keep that in mind. Plant pieces you buy in the store, and they'll usually grow. Further north, grow it in a pot and pull the pot inside during freezes. The next fall, dump out the pot and you'll have plenty of roots.

Guava

True tropical guavas will grow up into Zone 9 and occasionally freeze down in the winter. I tried growing pink and white types in my North Florida food forest, and they would freeze all the way back in the winter and fail to come back in time to bear anything the next year, so I gave up and grew guavas in the greenhouse... until I discovered a much more cold-tolerant variety known as the "cattley" or "strawberry" guava. This plant has become a major invasive pest in Hawaii but it's quite well behaved in cooler climes.

The flavor of cattley guava is good and distinctively guava in its musky sweetness. Cattley guavas are not super cold-hardy but an established plant can take temperatures down into the low twenties without dying. I grew some at the edges of oak trees, next to the house, and in somewhat sheltered locations, and they did great. Much better than the tropical types.

If you'd like a guava that's even more cold-hardy, check out the pineapple guava (*Acca sellowiana*), also known as feijoa. Though it's not a "true" guava, it bears a tasty green fruit and can survive cold down into the mid-teens. Mine never suffered any frost damage despite some cold winters by Florida standards. They are also attractive landscape shrubs. On the down side, they take a long time to grow big, unlike their fast-growing tropical cousin.

Jaboticaba

The jaboticaba tree is one of the most beautiful, exotic, and delicious trees on the planet. It has graceful and delicately turning branches with lovely flaking bark. The blooms are little powder puffs that appear right on the trunk and branches and then develop into purple-black grape-like round fruit. Imagine someone took a crepe myrtle that wasn't in bloom and glued purple gumballs all over its smooth trunk. That's how a jaboticaba looks.

Originating in Brazil, the jaboticaba is a small to medium sized tree which grows slowly and takes some years to bear. Once it does start blooming, it can fruit as many as six times a year.

So what about cold tolerance?

According to the California Rare Fruit Growers:

> *Some plants can take 24° F or lower and survive; others are damaged at 27° F. In 1917, a young tree at Brooksville, Florida survived a temperature drop to 18° F. with only the foliage and branches killed back. In California jaboticabas have been successfully grown in San Diego, Spring Valley, Bostonia, Encinitas, South Los Angeles and as far north as the San Jose and San Francisco Bay areas.[8]*

I never dared plant a jaboticaba out in my yard for a few reasons.

One, they are expensive and often rare trees.

Two, they grow slowly.

Planting a rare and expensive slow-growing tree in the yard as a zone-pushing experiment, knowing I'd have to protect it for years and then might still lose it as an adult... I just couldn't do it. Instead I put my tree in a big pot and kept it in the greenhouse. In retrospect, I was rather a sissy. You should totally blow the money on a few and plant them at the very edge of their range or beyond.

For science. Do it for science.

Jackfruit

If you subscribe to my popular YouTube channel, you may have seen the video I posted on my friends Chuck and Sarah who are growing jackfruit in their South Florida front yard.

The tree is almost, but not quite, as beautiful as Sarah. She's totally gonna kill me for saying that... fortunately I'm a zillion miles away from South Florida right now. Ha ha! Come and get me!

Their jackfruit tree is a truly magnificent specimen, loaded with highly valuable fruit. This is great, as South Florida is near the top of jackfruit's growing range and may actually be somewhat above it.

A mature jackfruit tree, though a tropical through and through, has the ability to survive temperatures into the upper 20s for brief periods. That means you can just barely grow them along the coast up into the Palm Beach area, with special care given to the trees when young.

According to the University of Florida: "Jackfruit leaves may be damaged at 32°F (0°C), branches at 30°F (-1°C), and branches and trees may be killed at 28° (-2°C)."[9]

I wanted to plant a jackfruit in The Great South Florida Food Forest Project I planted in my parents' Ft. Lauderdale yard, and, in fact, I did put a seedling in there at one point... but after seeing the size of Chuck and Sarah's tree, Mom vetoed the idea.

And the seedling *died mysteriously*.

That little yard would be great with a massive tree canopy and gigantic fruits no one in the house would eat, don't you think? C'mon, Mom! It's just like another mulberry tree! You can never have enough mulberry trees!

If you're way out of the range for growing jackfruit, as many of you are, then this portion may just be a curiosity... but if you're in the range where they grow, or even close enough to zone-push a bit, let me make the case for this marvelous tree.

In much of the tropics jackfruit is well known by many as a healthy, easy-to-grow delicacy that stands in as a starch crop, a fruit crop, and a nut.

In the USA, it's more of a "*eww...* that's weird" kind of a fruit. This attitude isn't helped by the fruit's strange smell when ripe.

Let's get the couple of negatives out of the way first.

Some has described the smell of ripe jackfruit as "boiled onions"; however, not all the jackfruit I've encountered smell like that. Some just smell somewhat fruity. Fortunately, once opened, there's no strange scent to the delicious flesh inside.

Another objection to growing jackfruit trees is that most people have never tried them and are afraid of growing something they haven't tasted.

That's a sissy excuse. Just go for it. Sissy.

A final reason people don't like jackfruit is the incredibly sticky latex in the rind and around the edible portions of the fruit.

As a counterstrike, here are some good reasons to grow jackfruit.

Reason #1: Jackfruit trees are beautiful.

The jackfruit tree is an evergreen, nicely shaped tree that is great for shade and as a specimen. In fruit, it's the kind of tree that causes car accidents as rubberneckers swivel around to see the huge fruit growing directly from its trunk and branches.

Not enough? How about some cold, hard capitalism?

STRANGE AND MARVELOUS CROPS TO TRY 101

Reason #2: Jackfruits are highly valuable.
In the right ethnic markets, you can get mad money for good jackfruit. I've heard of them selling for $10 per pound! That's pretty awesome, considering how many pounds one of these fruit can reach. Even if you never ate a single fruit yourself, you could likely cover one bill per month just selling jackfruit. A lot of immigrants miss the jackfruit of their home countries and don't have the space or the time to grow their own. Meet the need and profit!

Reason #3: Jackfruit are productive.
Jackfruit trees can produce hundreds of pounds of fruit per year and they will often produce for about half the year... and sometimes will have ripening fruit here and there year-round. This productivity happens with very little care. Jackfruit can even start fruiting from seed under ideal conditions in just a couple of years.

My friend Chuck harvested about a dozen just when I was visiting their home to record my video.

That was probably 140 pounds worth... with easily another 1,000,000,000,000 pounds on the tree!

Okay, I'm exaggerating—but the yields are impressive.

Reason #4: Jackfruit are delicious.
You know Juicy Fruit™ gum? That flavor was based on jackfruit. The story goes that the owner of Wrigley's gum tried a jackfruit and was blown away, so he took a fruit to his lab guys and said "make gum that tastes like this!"

The flavor of jackfruit is eminently tropical with undertones of passionfruit, pineapple, and guava. It's amazing. I can't imagine anyone not liking it.

Reason #5: Jackfruit is also a starch and a "nut."
The seeds of a jackfruit can be boiled and eaten like chestnuts or boiled peanuts—and, like its close cousin, the breadfruit, an entire unripe jackfruit can be skinned and roasted as a starchy vegetable. It's not just a sweet fruit—it's also a potential staple starch crop!

If you can grow a jackfruit tree, do it. Even if you never ate a single fruit, you'd be the wonder of your neighborhood.

Chuck and Sarah's tree is quite a looker—and they had no idea what the tree was going to be like when mature. They went out on a limb (*heheh*) and now are harvesting huge fruits they can share with friends or sell at local markets. Growing jackfruit is delicious, fun, and potentially profitable.

But what if you're not even near the edge of jackfruit's growing range? It would be quite tough to pull off growing jackfruit much farther north since it's definitely a tropical tree, but I'll give you my suggestions.

Growing jackfruit in frostier areas may be possible if you have an almost impossibly good microclimate or a greenhouse with some *serious* headroom! Remember the value of these suckers? It might be worth building a greenhouse just to grow them.

My friend Craig Hepworth grew a jackfruit tree inside his greenhouse over a pit in Citra, Florida, though it didn't fruit much, as it was pushing the ceiling and getting burned back now and again.

Build bigger and taller. Plant right in the ground, and prune like mad!

Jackfruit trees aren't a small tree and they don't like being constrained in pots; however, they really are easy to start from seed, so you don't have to spend much of anything if you want to experiment with growing them outside their range. Only fresh jackfruit seeds will germinate. They look like a large bean. Throw some in your compost pile, and they're likely to come up all on their own.

Another possibility is to be really clever with your south-facing wall. Since I grew coffee, key limes, guava, black pepper, and lemons next to my south wall in North Florida... why not try a jackfruit? Judicious pruning, tying down branches and keeping the trees tight to the wall might work; but still, those trees get big! I've never heard of anyone espaliering a jackfruit; you could be the first to do so! If you have a two-story house it would really help.

I might start the tree from seed and let it grow straight up planted toward one end of the wall, then bend it sideways as far as I could once it got big. Or start it in a big pot until it grows to about six feet, then plant it way bent over along the ground. Weighting it down with cinderblocks might work, too.

On the problem with headroom in a greenhouse... there may be another solution.

The book *Agricultural Options for Small-Scale Farmers* reports:

The jackfruit usually bears its large fruit (up to 80 cm in length) on the trunk and main branches, high up in the tree. In the Songkhla Province of Thailand, the young jackfruit is planted over a large stone or metallic plate, thus blocking the growth of the tree's tap root. As a result, fruits grow in clusters around the base of the trunk.[10]

This would be very helpful in a small space!

Once a tree reaches up past the edge of the roof, that part is going to freeze off on a cold night. Your goal is to dwarf the tree and keep it from getting that tall.

Another possibility might be to grow a jackfruit tree on an island in the middle of a pond. Or on a big raft in your swimming pool. The radiant heat overnight could keep it alive. Maybe. Or maybe that's just crazy talk.

I planted a jackfruit seedling in my North Florida food forest once, and it froze to the ground. The next year it came back from the roots, then froze down again that winter and failed to return.

Growing jackfruit much outside its range is a serious challenge. I'd never recommend it to a beginner, but it's totally something I would try myself until I pulled it off.

If anyone tries it, please let me know what happens!

Kangkong

Kangkong (*Ipomoea aquatica*) is a tropical Asian vegetable that likes swampy areas and ponds. A relative of morning glories (and sharing their pretty flowers), it's remarkably productive. The leaves and tender young stems are quite delicious when lightly steamed or sautéed.

Though it cannot take any frost, I managed to keep kangkong alive outdoors in North Florida through the winter during the years that my pond didn't freeze over. If it had gotten much colder, though, it would have melted. This vegetable is perennial if you can keep it alive during the winter. I always kept a backup patch of kangkong growing in my greenhouse. Because the stems will root anywhere, it's easy to just take a little piece (or many pieces) from outdoors before winter and pop it into your fish tank or a jar on your windowsill to root. Once the warmth returns and the danger of frost has passed, plant it outside again.

Kangkong grows easily in bathtubs, kiddie pools, and other unlikely places. It will also grow in the ground as long as it isn't too dry; however, the stems seem to be a lot tougher and less palatable than ones grown in the water.

If you live in a more tropical area watch out, as kangkong is an aggressive invader in warm waterways and can be a real pain-in-the-neck invasive when it escapes.

Lemongrass

Lemongrass is a remarkable grass. To me the flavor is reminiscent of citrus and Froot Loops®, though it's less likely to turn milk weird colors. It definitely has some lemon flavor and is beloved in Southeast Asia for its unique flavor. Like its relative citronella grass, lemongrass repels insects. Some plant it to reduce the amount of mosquitoes hanging around their gardens but I've found that smoking a huge cigar works better.

Lemongrass is a tropical grass and will freeze to death if the ground freezes; however, I've had luck growing it outside of its range by harvesting it earlier in the year and letting it get tall and bushy in the fall... then leaving it alone! If you let it stay a nice big clump of grass, it self-insulates to a certain extent. The top dies but the resultant drooping mess of leaves helps blanket the core of the plant and keep it alive until spring. I have lost some younger plants to the cold. Mulching will help, as will keeping them growing in a sheltered location or beneath a bit of tree cover. The one I grew beneath my yaupon holly on the north side of my house came back faithfully year after year, providing us with as much lemongrass as we wanted.

Bonus: the oil in lemongrass leaves enables the dried blades to be used as a very good fire starter. How crazy is that?

Loquat

Loquat trees are more cold-hardy than many of the plants in this book, so much so that they're technically a sub-tropical and not a full-on tropical fruit. Some sources claim they'll go through a 10-degree night without damage.

The problem, as I mentioned previously, is that they bloom at the wrong time of year and the blooms are not hardy below

27 degrees. This means you could grow loquat trees in, say, Oklahoma... but you wouldn't get any fruit.

If you want fruit, you need to tuck the trees into a sheltered location or somehow keep the blooms from freezing on cold nights. And that's not all: you'll also have to protect the fruit, which usually ripens in February and March. Not good.

My friend Oliver Moore in North Florida has a variety of loquat he calls "Shambala" that exhibits a longer bloom cycle, often ripening some fruit later in the year after frosts; however, I'm not sure if this variety is available anywhere yet.

Your best bet up north is to plant in a sheltered location for fruit and/or pray global warming is true.

Mango

As a child I greatly enjoyed eating the frozen mango slices my grandmother would store up in her chest freezer from the tree out back. When that tree produced, it produced in quantity, and she was diligent about slicing and preserving the fruit so we could enjoy it nearly year round.

In South Florida where my grandparents lived, there were mangoes all over the place (provided various immigrants didn't steal them from your trees), but as I learned when I moved to Tennessee, good mangos are really hard to find outside this tropical tree's climate range.

Despite the claims of "cold-hardy" mangoes, I have yet to read credible reports of mangoes that will sail through frosts without damage. This makes your job tough.

In Frostproof, Florida (which isn't really frost proof, despite the charming name), a retired missionary friend of mine planted a seed from my grandparents' mango tree years ago and grew it into a large tree that has lived through multiple frosts but has always failed to fruit, despite its impressive size.

If I were to attempt growing mangoes outside of their range without a greenhouse, I would plant a "condo" or "patio" mango by a south-facing wall and keep it pruned short. These types are dwarfs, unlike typical mango trees which can reach a height of over 100 feet in the tropics. A short tree, kept close to the wall, is likely to live even a hundred or more miles north of its comfortable range. I have found that even large mango tree varieties can be brutally pruned and will fruit in pots. One I grew from seed fruited in a pot at less than 6 feet tall after four years. Exciting!

Miracle Fruit

The miracle fruit (*Synsepalum dulcificum*) is a strange little horticultural curiosity. The fruits are nondescript in flavor and somewhat sweet; however, once you eat one, it will make anything you eat after it taste amazingly sweet. My friend Craig with the awesome greenhouse shared some that he grew with me and then we chased them with lemons and starfruit. Both tasted like sweet candy afterward. It's quite a strange sensation. Even vinegar will taste sweet, though when you swallow it will still burn your throat like drinking acid. Yeah, I tried it. I won't do that again.

Miracle fruit don't like cold at all, but they are easy to keep in a container. The big thing is to make sure they stay watered and are growing in acid soil. They'll fruit even as tiny plants once they're established, though they grow slowly. Fruits can appear year round, but they like warmth and humidity and don't seem to like full sun. Since the cold messes them up quickly, try a south-facing wall or a greenhouse.

Moringa

Moringa trees (*Moringa oleifera*) have gained serious popularity in recent years because of their health benefits and ease of growth. They have edible leaves and pods which are very nutritious. I grew them in North Florida despite their truly tropical nature after discovering from friends that they'd come back from the roots even after losing the entire top of the tree. I didn't get any pods to eat, but I did harvest lots and lots of leaves every year. We added these highly nutritious greens to soups, scrambled eggs, and a wider variety of meals. When we were sick, we also brewed them into tea via boiling the green or dried leaves.

Moringa grows very quickly and can reach 10 to 20 feet in height over the summer... from seed! Yet one night below 32 degrees and the woody trunks of moringa will turn into mushy rotten columns.

If you let your moringa trees freeze to the ground then grow back again from the roots sometime in spring, you'll be waiting on new leaves for a lot longer than necessary, and your yields will be much lower than if you protect a portion of the trunks.

Protecting moringa trees from frost damage is the number one thing you can do to ensure an early spring harvest of nutritious young leaves and that you get big, healthy trees that last long term. Here's how I do it.

First chop your moringa trees down to 3- or 4-foot trunks in late fall or early winter. I wait until the first frost is coming, then do this the day before. It hurts to cut the trees down, but you can take away some of the pain by drying the tree's leaves to use through the winter. I usually put away a couple of dry gallons of leaves... that's a *lot* of moringa. We never run out.

Now take some old chicken wire or other fencing and make some wire rings to put around the tree, leaving about a foot of clearance between the trunk and the wire on all sides of the tree.

I usually make my rings about 16 to 20 inches across, depending on the size of the tree. Now you just need to stuff them with some nice, dry insulating material.

We used to buy straw for this step, but, unfortunately, I have discovered that a lot of straw and hay is now contaminated with deadly long-term herbicides (aminopyralid and clopyralid being the most egregious examples) that do not break down quickly. As the straw gets wet and rots, it can poison and slowly kill your moringa trees.

I have been warning gardeners about this problem for years and have been sent many pictures of wrecked gardens and poisoned plants since sharing it on my site and in *Compost Everything*. It's really a shame since straw in particular is a great material for mulch and compost as well as being useful for holding in heat during the winter. Long ago I got permission to sweep up the straw and hay at Lowes and the local feed store. I would pile it everywhere in the fall, using it to protect the roots of my perennials in Tennessee and as great fodder for the compost pile. Now I don't dare, as the risks are too high. Once I got hit with a batch of contaminated manure containing aminopyralid-laced hay consumed by cows, and it destroyed my spring gardens. I then dug deeper and found the problem is rampant. My friend Andi also suffered serious damage to her food forest after mulching with rotten hay from a local farm. It's not worth rolling the dice!

Now I usually use leaves instead. Another year I used pine needles I raked up in a neighbor's yard. A reader warned me that the "pine straw" sold in bales is often gathered from areas sprayed with herbicide to remove weeds, so I can't even recommend that formerly safe material.

To protect your moringa trunks, stuff the wire rings with leaves, then cover the tops of the rings with something. A piece of cardboard, a chunk of tarp, a big trash bag... whatever. You just don't want the leaves getting soaked completely and causing the trunks of your trees to rot while the moringas are fitfully sleeping through the cold of winter.

In cold weather moringa trees won't grow but so long as the freezing weather doesn't get to them, they'll stay in stasis until the weather warms. After all danger of frost has passed, unwrap the rings, rake away the leaves and stand back. Moringa grows back at an incredible rate when it has plenty of trunk left to support it. They can take repeated and brutal pruning, so don't worry that hacking the tree down to 3 to 4 feet will kill it. It won't.

They survive at a higher rate this way than if you let the cold take the whole top of the tree, too. Generally, they'll come back from the roots in that case but not always. I've never lost them when they're dry and frost-protected by rings of wire and leaves. I did some side-by-side testing with preserving the above-ground stump of trunk and with just letting others freeze. The size of the trees and their leaf yield were quite remarkable.

I would venture to say you could grow moringa trees up into Georgia with this method. The species is very tough, provided it doesn't get frozen. It only takes a few minutes to do, and the results are excellent.

If you live further north, moringa can be grown from seed as an annual vegetable and will still give you good yields before frost. It's a very fast-growing tree... startlingly so.

Papaya

Papaya can be grown into zone 8 if you know the right tricks. I was told it's "impossible" to do in North Florida's climate, but

STRANGE AND MARVELOUS CROPS TO TRY 113

nevertheless, in the fall, we usually ate papaya for breakfast on a regular basis. They're really good.

A gardener visited me one day and remarked "What, you have papaya? Too bad you can't grow those here!" They were serious and were looking at the trees when they said it. "Impossible!" I had to laugh.

Papaya, as you know, is a tropical fruit. It's not a real tree; it's actually just a big fleshy plant that develops a woody stem over time. Think of it as a giant melon-producing herb.

If it gets cold, the tree will rapidly wilt and suffer. A hard freeze that lasts any period of time will kill a papaya tree right to the ground.

The first year I grew papaya, they had their tops whacked off by a frost despite being wrapped in Christmas lights and planted pretty close to a south-facing wall.

Fortunately, papaya trees grow very quickly and bear fruit rapidly, even from seed. If you want to eat papaya, I find that it really helps to start your trees one spring in pots, then plant them out in the spring of the next year.

You don't need to buy papaya trees from a nursery in order to try zone-pushing them. All you need to do is buy a papaya from the store and plant the seeds.

Know this: papaya seeds do not keep very well long-term. They quickly lose viability. I've found the best way to grow papaya from seed is to scoop out a handful of seeds from the middle of a ripe fruit, slop and all, and then mush that sloppy handful right into the soil. I'll fill a few pots with potting soil and chuck in wet seeds in handfuls, crumble the soil around, and keep them well watered in a warm place until the seedlings emerge. As they emerge, I use scissors to thin them out, removing smaller and weaker sprouts and selecting the ones that look best to me.

I do this in multiple rounds. A pot may start with 25 seedlings in it... and I thin them down to the best eight or so. When those get about 8 to 12 inches tall, I pot them in their own pots and let them fly. Papaya like full sun and good fertilizer to reach their potential. If they're in hot dry sand without enough water they'll usually live but will fail to bear much fruit. When I planted them with lots of water, compost, and thick mulch, they bore fruit within a year from seed.

If you start papaya in pots in spring, get them good and big, and then bring them in on cold nights or pop them into a greenhouse when fall and winter arrive.

In spring when all danger of frost has passed, plant your papaya trees in rich soil in a somewhat sheltered location. Against a south wall is great—just plant them as close as you can. RIGHT AGAINST the wall is best. Growing dwarf varieties of papaya makes them easier to protect; however, even if you plant your little trees in the garden in spring with lots of water and good compost and on top of something horrid and nitrogenous, they're going to fruit. It's also important to plant at least a few papaya for the sake of pollination since papaya come in male, female, and hermaphroditic forms.

In a rich garden bed, papaya will grow like gangbusters in the heat and can start bearing fruit by late summer. By fall, that fruit will be ripening. If a frost threatens, pick the fruit and bring it indoors to ripen, or cover the trees with blankets and pray over them.

Another method is just to grow the papaya trees in big pots and haul them into your greenhouse for the winter. They'll keep producing for a few years this way.

Growing papaya may not be the most efficient way to spend your gardening time, but the flavor of ripe papaya along with their health benefits makes them a great addition to your homestead.

If you end up with a bunch of green papayas right before a frost and they're too green to eat, you can use them as a starchy vegetable in various recipes. The Thai people also make a very good raw green papaya salad (som tam) with dried shrimp and chili peppers.

I really, really like papaya. So much so that I thought of building a separate greenhouse just for them and then selling the "impossible" fruit at a local farmer's market, far from their tropical home.

If you grow papaya by a south wall and the tops freeze off, don't worry too much. In the spring they'll often grow back again and bear another crop by fall. I once visited a foreclosure in North Florida and saw a gnarled papaya tree growing which had obviously frozen down and regrown again for years. It might not bear fruit thanks to its open location without cover, but it didn't die... and that's a start.

If I had followed the standard advice, I would've never planted papayas at all. Now that I've pulled them off multiple times over multiple years, I know "common sense" isn't always correct.

Perennial Cucumbers

Perennial cucumbers are a real thing. Most gardeners don't know anything about this delicious and productive vegetable, but in South India they're a common fruit and cooked as a vegetable in curries. The Latin name for the perennial cucumber to which I refer is *Coccinia grandis*, also known as the ivy gourd. The Hindi name is tindora. It's classified as a noxious weed in chunks of the tropics, but many of the vines growing in colder climes seem to be sterile or, at the very least, non-aggressive forms.

Perennial cucumber fruit are about 2 to 3 inches long when ready to eat. They ripen to a scarlet red but at that point it's too late to harvest unless you're saving seeds. The flavor of the smooth green cucumbers is slightly sweet and tart, almost like a dill pickle right off the vine with a little bit of an astringent undertone like many other cucumbers.

I first encountered this tasty vegetable as a child. Our neighbors across the street were from India and the wife had a variety of beautiful and edible exotic plants growing in her yard. Sugar apples, curry tree, super-hot peppers... and then on one fence, a big mess of perennial cucumber vines. She let us help ourselves to the fruit, and they were quite good. I always loved cucumbers but never realized there was a type that could live for years without replanting. The cucumbers I grew in my own garden were good, but the harvest season was short, and they often got knocked down by pests.

After our neighbors moved away and I moved to my own home, I occasionally remembered those perennial cucumbers and wondered where I could get some of my own to grow. Years later I was at a gas station run by Indians somewhere in North Florida, and I saw a pile of little cucumbers on the countertop for sale. I knew the manager since I had provided him twice with cassava cuttings after talking gardening one day.

"Where can I get seeds or plants for those?" I asked.

My "friend" behind the counter looked at me suspiciously. "You can buy the fruit from me."

"I want to grow my own," I replied.

He shrugged.

I was irritated but wasn't about to give up. I noted the fact that some vines were growing on the fence out back but I wasn't about to steal a piece for my gardens. That was definitely crossing a line.

I waited and watched, then drove by the gas station again the next spring with a packet of moringa seeds I'd harvested from my backyard moringa orchard.

Inside there was a woman at the counter. No sign of my plant-hoarding nemesis.

"Hi," I said, "I'm a gardener, and I shared some cassava cuttings with the manager one time. Here are some seeds, too."

"Thank you," she said, looking at me somewhat suspiciously. I was getting used to this.

"So," I queried, "how are your vines going? The little cucumber vines? I would love to take a few cuttings!"

She shrugged. "They are all dead. The winter killed them all."

"All of them?" I replied.

"Yes, I have none to share."

"Thanks," I said, "I'll go take a look. Have a great day."

Before she could object I walked out to the fence. She was still inside. There along the ground was a great number of little sprouts rising from the soil. Ivy gourds everywhere.

I carefully pulled one with a few roots and then looked back at the gas station. She was now outside, frowning, talking on the phone. She started to yell something but I was quicker.

"Thanks! I found one! Isn't this great? Thanks so much!" I yelled, then fled to my car.

I had done it. I was now the proud owner of a perennial little cucumber plant I didn't exactly steal because it was dead and didn't exist anymore! She had said so herself!

Growing perennial cucumbers is easier than growing annual cucumbers. They have less disease problems and also have a tuberous root beneath the soil which can sail through winter even if the entire top of the plant is whacked down by frost.

Some people in colder climates will dig up the root in fall and pot it, then plant it out again in the spring. You can also take cuttings before frost arrives. They root easily, and then can be planted in spring.

Perennial Peppers

Though bell peppers are almost invariable annual, many species of hot pepper can be kept alive for multiple years if they aren't killed by frost.

I saw a 6 foot tall cayenne pepper bush in Ft. Lauderdale once growing in the backyard of a Puerto Rican friend's house. He'd had it there for years, and it was loaded with hot peppers. I also grew a habanero pepper into a nice specimen in a pot for about three years before forgetting it outside on a frost night. It wasn't hot enough to beat the cold, Scoville Units notwithstanding.

Many hot peppers, though they fruit quickly and are grown like a vegetable, are really woody shrubs that will happily keep bearing for a long time if you do your part to ensure their success.

I decided to mulch around some of my cayenne pepper plants before winter one year to see if they would come back from their stumps the next spring. Most of them died but at least one came back again. Instead of giving up, I decided the next year to grow multiple hot pepper plants in pots and then bring them inside on cold nights. That worked! Another thing that works to keep them growing was planting by a south wall. Peppers like a lot of sun, though, so if you try to protect them beneath a tree canopy their yields will greatly decrease or cease altogether. Keep them hot, and they'll provide you with lots of heat.

They grow well in pots and can live for years even up north if you pull them into a greenhouse during the cold.

Pineapples

Hawaiian gardeners have all the luck. Roughly 1.5 zillion amazing tropical species grow there, including pineapples. On most of the continent, though, life is hard for gardeners.

You'll do great with pineapples until the temperatures drop into the 20s, then your pineapple plants melt. The cold is insurmountable.

Fortunately, they're easy to grow in pots. I inherited a bunch of pineapple plants from my grandfather when he passed away a couple years back. He was in South Florida, and I was in North Florida. This presented a problem, as the frosts in North Florida would take out pineapples. You might pull them off for a few years (one farmer did this a hundred years ago in North Florida) but eventually there will be a bad winter (which ended the work of that enterprising farmer).

Before I go further, I have to tell you about my grandfather. His name was Judson Greene, and he was a sailor and a brilliant carpenter. When I was a kid, he stood ten foot tall. If you dropped him on a desert island, he could build a sloop from laminated twigs, find a way to varnish it (twice), then sail back to home port before dinner. He had traveled about the tropics in the Navy, done scientific research, and still managed to raise five children, the oldest of whom was my mother.

Grandpa was an inveterate experimenter, which is perhaps where I get my own exploratory drive. He planted a rubber tree in his backyard, grew a mahogany in the side yard, and was always toying with the idea of converting his whole house to solar power... before that was anywhere near feasible or common.

When he discovered in his 70s that you could grow pineapple plants from the tops of store-bought pineapples, he hatched an idea. Visiting the local grocery store, he asked the woman behind the food prep counter what she did with the pineapple tops that were removed when they made fruit salads. When he found out they were being thrown away, he asked if she'd save him some. She did—an entire trash bag full.

Grandpa called my younger sisters and their friends together and made them an offer: for each pineapple top they planted in his backyard, he'd pay them a nickel.

Pineapple plants were rapidly planted here and there all through the backyard landscaping. About two years later, the bounty started trickling in, and wow—those pineapples were the most delicious golden fruits. The ones the raccoons didn't steal were shared regularly with visitors, who were all amazed to realize they were homegrown. For the next decade and until his death, Grandpa had pineapples... and his army of young planters grew up enjoying the fruit.

Though you may not live in the right climate to grow pineapples unprotected, they're remarkably easy to grow—even outside the tropics.

When I inherited as many of Grandpa's as I cared to dig up, I took them to my homestead, potted some and planted others up against the south wall of my house. The frosts only claimed a few test pineapple tops I later put out in the yard.

The key to growing pineapples is two-fold.

1. Don't let them freeze!
2. Don't overwater them!

That's the basic formula. You can plant pineapple tops in cheap potting soil and water them as you remember, and they'll grow. Pineapples, like all the bromeliads I've ever handled, have limited root systems. They feed primarily through their leaves. That said, don't ever let any fertilizer fall down into the center of the plant. That can burn the poor thing and rot it from the middle out.

If the plant is looking really yellow or the leaves are getting washed out and reddish, it needs a little fertilizer boost. I use heavily diluted urine, compost, a little dissolved Epsom salt, and fish emulsion every once in a while, but they probably don't even need that.

And for those of you who live for ornamentals—why grow other bromeliads when you can grow pineapples? They're attractive and edible. The downside is that they take a couple years to produce from a top or a slip, but the upside is that you can start them any time you want, put them aside, start more, put them aside... and eventually, you'll have tons of pineapples. Just do it in between taking care of your faster-producing plants, and you'll get there.

If you're farther north than me and even the south wall of your house is too cold for growing pineapples, bring the pots in during the cold, and place them next to a sunny window. They'll live. In fact, I've even seen small varieties used as novelty houseplants.

Here's another story about pineapples. I was once helping to paint a house in South Florida (I have managed to paint quite a few houses—isn't it bizarre that a writer might need supplementary income on occasion?) and I saw something strange. The elderly owner was peeling an apple onto one of her pineapple plants out front.

"What are you doing?" I asked.

She looked up, her eyes shot with keen fire like a hawk weighing me in the balance and finding me lacking in what was obviously common knowledge. Eventually she answered, "I'm making them fruit."

I didn't dare ask any more. She was really intense, and I knew she was quite smart. I was younger, and, somewhere in my subconscious, I may have been somewhat afraid she would peel me next.

Years later I discovered that ripening apples produce ethylene gas which triggers blooming in pineapples. It sounds crazy but it's true. You can try this trick yourself. Just do it with pure faith or it may not work.

Though you may not be able to grow a little plantation like Grandpa did, you can certainly pot up a couple of pineapples for your porch. As tropical plants go, they're one of the easiest to grow and the easiest to find locally.

Try it! When you're sipping a homemade pineapple cocktail, watching your neighbor hoe their oh-so-boring vegetables, you'll thank me.

Soap nuts

If you're a back-to-the-land sort or an alternative health, organic-market-shopping type, or a plant lover, or even a prepper, you may have come across soap nuts before. If you've bought them in the store, you've probably bought the "nuts" of the tropical Indian tree *Sapindus mukorossi*. Though this tree is allegedly a purely tropical species intolerant of frost, it has a cousin native to the U.S. which can handle quite a bit more cold. Meet *Sapindus saponaria*, also known as the Florida soapberry.

I've been told by the highly knowledgeable author Dave Chiappini of Chiappini Farm Native Nursery that the only common range of Florida soapberry is a tiny geographical area scattered across a few islands on the coast of Florida. Fortunately for us plant geeks, it's been saved from obscurity and is now being planted in a broader range of locations.

According to the University of Florida the Florida soapberry is hardy to USDA Growing Zone 10. This is demonstrably false since there are large specimens growing in Gainesville and bearing fruit quite happily right at the edge of USDA Zone 8.

This is great because this tree grows soap. Soap on a tree.

For preppers, homesteaders and cheapskates, this is good news.

The fruit, erroneously called a "soap nut," is loaded with saponins. Dry them (and pit them if you like), and they can be used to wash your hands or do a load of laundry when placed in a mesh bag. They last quite a few washings, too.

When I first started growing these graceful and useful trees, I read that the trees take eight years or longer to produce fruit when grown from seed; however, my friend Alex Ojeda reported that his soapberry trees bore fruit only three years after germination. Great news!

Germination is easy with soapberry trees. I scarified a bunch of seeds with a knife to increase germination and planted them in little pots. I had an almost 100% germination rate.

These trees do have a catch on cultivation: like many uncommon species which haven't been bred by man for generations, the soapberry needs a mate for pollination. Trees come in male, female, and hermaphroditic varieties. Only females and hermaphrodites will bear soap nuts. If you plant three trees, chances are really good that at least one or two of them will fruit for you.

I planted five in my yard because I wanted lots and lots of soap, plus I think the nuts are a valuable commodity for the farmer's market.

And really, won't the Econopocalypse be better when you can take a nice shower between bouts of fighting off diseased and drug-crazed looters with a broken shovel?

Soapberry trees grow tall with an airy, open habit. In fact, they look a lot like the despised Chinaberry tree that's invaded railroad tracks and roadsides across the south, though, unlike Chinaberry, they have almost white bark. They're quite attractive.

If you have a small yard, I recommend planting three in a tight triangle so they grow like a triple-trunked tree and pollinate each other without taking up too much space. That's what I did in my backyard, spacing them about six feet apart, although you could probably plant three in the same hole about 18 inches apart, and it would look really cool.

Growing soapberry is easy. Soap nut trees are tolerant of poor soil and grow rather quickly into lovely trees that don't cast particularly dense shade. Tuck some in on the edge of your food forest if your zone will take it. And if it won't, find a way to push the zone!

Just don't eat the fruit; the seeds are reputedly poisonous.

I don't eat soap anyhow, at least not since my legs grew long enough to outrun my mom.

Starfruit

Starfruit are a sweet-sour and quite refreshing fruit that grows on an attractive tree. They are not huge trees and will bear when only about six feet tall. Starfruit respond well to pruning, making them another good bet for espalier. In Florida they grow as far north as Orlando, although even there the frosts hurt them and will sometimes kill a tree. In ideal conditions they can bear fruit over long seasons or in a twice a year cycle of boom and bust.

Try the south wall thing. I've seen them grown next to a house in Gainesville, though they definitely suffered in production due to lots of shade overhead.

Sugarcane

I cover growing sugarcane in my book *Totally Crazy Easy Florida Gardening*, but I will repeat myself here, as this delicious grass is well-worth trying in colder climes.

There is a big sugarcane industry in the southern portion of Florida, yet I had great luck growing it in dry ground much farther north in the state. People have this idea that sugarcane is something that requires year-round tropical weather and a big old swamp. Fortunately, that idea is wrong. You can grow sugarcane successfully all the way up into Georgia, swamp or no swamp.

Other than its delicious flavor, sugarcane is also attractive as an ornamental. Depending on the variety, the thick canes can range in color from dark red-browns to yellow-green and have a very similar appearance to bamboo in the landscape. Since it's a perennial plant, once you plant sugarcane you can look forward to having it for years.

The hardest part about growing sugarcane might be finding

the plants in the first place. I've never seen it for sale at a plant nursery. Ask for sugarcane, and you're likely to get a blank look and the question "Does that even grow here?"

It's okay that they don't have any. You really don't need to buy a potted sugarcane plant. All you need is a good hunk of sugarcane with a couple of intact nodes (the joints in the cane). Since sugarcane is usually harvested in the fall, that's the time you're likely to see the canes for sale. Most grocery stores don't carry sugarcane, but a lot of farm stands do. I drove down 441 in North Florida one afternoon a few years ago and bought two different varieties of sugarcane from two different produce vendors located only a few miles apart. Grab a couple of stout canes (they're usually 5 to 6 feet long with about 8 to 12 nodes, depending on the cultivar), and you're well on your way. Try oriental markets and organic groceries with expensive produce sections if you can't find sugarcane at your local farmer's market or stand.

When you get home, cut your canes into segments with at least 3 or 4 nodes each, choose a sunny spot to plant them, then put those pieces on their sides about 4 to 6 inches down and cover them up well.

When I plant sugarcane in November, the plants always pop up for me sometime in March or April. For each cane you bury, you'll usually get a couple of good shoots emerging from the ground.

If you really don't want to trust the earth to take care of your little baby sugarcane plants, you can just stick some chunks of cane in pots with a node or two beneath the dirt and keep them someplace that doesn't freeze, like a sunroom. They'll grow.

When sugarcane shoots appear in the spring, and I'm pretty sure it's not going to freeze again, I fertilize them with chicken manure. You can also use lawn fertilizer if you're not picky about being organic. They're a grass, and, like lawns, they want lots of nitrogen. Throughout the summer they'll get nice and tall and sometime in July or August you'll really see the canes starting to thicken up, but don't chop them yet (unless you really can't stand to wait). Wait until it's just about time for the first frost of the fall or winter, then go cut the canes down, and you'll get the largest harvest possible.

If you don't cut them down and you get a freeze, you're going to lose all the above ground growth, and you may even lose the plants. Harvest by cutting the canes down close to the ground, and then put the sugarcane roots to bed for the winter by mulching over them with some rough material. Leaves are good for this, but probably any mulch would work fine. My sugarcane came back even when I barely mulched over the roots.

In its second year, sugarcane will bunch out and give you quite a few more canes than it did the first year, which means you'll be able to share the abundance. Or you can make sugarcane syrup, molasses, sugar, or my wife's favorite, rum.

Vanilla Orchid

Oh boy, now you're stretching! Growing vanilla is a tricky business even in its native tropical range. This isn't because the plants are hard to grow there, it's just that pollination, harvest, and preparation of the "beans" are all labor intensive.

I once kept a vanilla orchid growing over the bathtub by the filtered light of a window. They like warmth and humidity and are unlikely to survive any winter outside since most of the orchid is not in the soil but in the air. Cold breezes and cold nights will knock them out quickly. A friend kept a great big one going for years in a greenhouse where it sprawled over 35 feet along the edges of the structure; however, he didn't get any beans.

If you manage to grow vanilla beans outside their range, please tell me about it at TheSurvivalGardener.com. And share pictures! I would love to post any information you gather. It's a beautiful and exotic plant that many of us dream of growing.

Yams

Yams are not sweet potatoes. True yams are an unrelated group of species, some of which are staples in Asia, Africa, the Caribbean, and South America. Though generally considered a tropical crop, many yams can be grown in climates colder than you'd think. At least one species produces crops all the way north into Massachusetts.

Though you probably don't dream about growing yams like you might dream about growing coffee or chocolate, they really are a good root. I like them more than potatoes, and I like potatoes.

The reason you can grow yams outside the tropics is that they have a dormancy cycle that coincides with winter. Where I live in the tropics, they emerge in the spring, grow like crazy up trees, fences, and brush, and then die back in the fall to the ground. There are no frosts here; this is just how yams roll. In North Florida, they did the same thing, except that frosts would occasionally take down the vines a little early. Not to worry! The roots beneath the ground slept happily until warmer days arrived and then sprang back for the sky.

The exciting thing about yams is that they are a perennial that makes really big roots. This makes them a great survival crop for the Deep South, as well as being important in the tropics.

On many occasions I came across escaped edible yams in the woods of North Florida, dug up the roots, then took them home for my wife to cook.

Many yam species have aerial "bulbils" (roots) that you can plant for the next year's harvest. Some do not. Yams are only rarely grown from seeds except for breeding purposes, and if you live in the U.S., you're unlikely to have a long enough warm season or proper light cycles for them to even bloom.

The normal method of propagation is via bulbils for the varieties that produce bulbils, and via divided roots for those that do not.

If you don't have bulbils, you need to make "minisetts," which for some reason is the technical term for chopped up chunks of yam root. I dunno why since I don't get to make up the names. All that is required to make minisetts is a good yam root, a knife, and perhaps some ashes to ward off potential soil pests.

Getting a yam root is easier than you'd think. Just go to a local ethnic market and hunt one down. They're usually called "name" roots, pronounced *naw-may*.

Cut your yam root into chunks about the size of a peach, dip them in ashes, then plant them.

I put a bunch of ash-dusted yam chunks into a big pot full of dirt in early spring, then transplant the ones that sprout into my gardens and food forest.

Not all of your yam minisetts will grow; however, most will root and give you some yields.

Another method I haven't read much about is starting yams from cuttings. I had good luck rooting yam cuttings in a mist house. It was surprising how easily the yam cuttings rooted. I don't know if they will give you as big of a harvest the first year if you start them from cuttings, but I do know they'll root.

Yams need something to climb. They grow vigorous vines and will happily shoot to the top of a tree if given half a chance.

I plant mine just under the surface of the soil near something—anything—they can climb when they emerge. I've grown yams on fences, on trellises, on an unused clothesline and even on a pollarded sweet gum tree I used as a living trellis. ("Pollarding" a tree means to prune it back repeatedly at the same height, eventually causing large, flat knobby growth where you made the cuts and keeping it from getting too tall. Quite handy for small yards—or for making living yam trellises.)

If you have bulbils or minisetts available, plant them in fall, winter, or early spring.

Remember: yams have a growing season and a dormant season. Summer is their most active time of growth. They grow vigorously through late spring and summer, slow down a bit in the fall, then die back and eventually freeze down in the winter.

As the growing season progresses, yams start making their bulbils (if they're a yam that does that) which mature in the fall. The below-ground root really seems to do a lot of its growing into the fall as well, preparing for the winter ahead.

Some species are grown JUST for their bulbils, such as the rare edible forms of *Dioscorea bulbifera, but you're unlike to score those varieties anywhere and the wild forms (and the ones sold on ebay) are almost all poisonous. Not good.

My best luck has been with its non-toxic cousin *Dioscorea alata*, also called "the water yam." It's a big, gnarly thing that grows like a beast. I've dug 40lb roots before. This is the one I scavenge from the woods, not the more common and usually toxic *D. bulbifera* "air potato" of invasive species fame.

Growing these yams and the name types is possible through a chunk of Texas, Arizona, Georgia and other places not known for yams. I would love to move further north at some point and see how far I could push their growing range.

Maybe trellises on a south wall? Perhaps a nice, sheltered food forest with yam vines sprawling over apple and pear trees?

If you live in zone 6 or 7, there's still a yam for you. The cold-tolerant Chinese yam (*Dioscorea opposita*) can be grown for both its large underground roots and its tiny little edible bulbils. Eric Toensmeier covers this species in his book *Paradise Lot*, calling the bulbils "yamberries" and remarking on their prolific nature. His book recounts the story of a permaculture garden paradise he created in a blighted neighborhood in Holyoke, Massachusetts, so yes, there are cold-climate yams!

Yams don't need a lot of care or watering to stay alive, though taking care of them will raise your yields and reduce the time needed until harvest. The yams I grew in great garden soil with lots of compost and water made big roots in their first year; the ones I grew without any care whatsoever generally took two years to make big roots. Wait more than two years to harvest, and you're likely to have portions of the roots that are woody, withered, or otherwise unfit for the table... though the new growth will still be yammy. I mean yummy. Actually I mean both.

Since yams are a perennial crop, you can simply plant them one year and then dig them up a year or three later when you're hungry. There's going to be something worth eating there.

I cook yams just like white potatoes, though I find they cook faster and brown up nicer than potatoes do. Once you know how to grow yams, you'll be eating them all the time.

Yams also make wonderful roots for the Crock-Pot® and really good French fries. I also really like them shredded with a cheese grater and fried into hash browns. And bacon. Just make sure you cook them through as they contain some oxalic acid when raw and that will make your throat and lips feel scratchy. Some people also have issues with skin irritation while peeling yams—they bother my wife but not me. Rubber gloves are a good fix if you have sensitive skin.

Yams keep pretty well on the counter. Unlike potatoes, you don't have to worry about them greening up and poisoning you. If you store them under moist conditions, they'll start growing lots of little roots and new buds. I left some in a plastic bag once, and they did just that, so I ended up chopping them up and planting them instead of putting them on the table for dinner.

The best place to keep yams is right in the ground, then you can dig and eat them as needed.

If you have a great big root, you can actually break or cut pieces off of it and the cuts will dry up pretty well without ruining the rest of the root.

This is good when you have a forty-pound monster to consume.

Conclusion

These are just a few of the crops you can attempt when pushing the zone. I'm sure there are dozens, or even hundreds, more you can imagine. Once the door is opened to the tropics, the experimentation and the fun begin.

Appendix A: The Plant Hardiness Zone Map

The USDA Growing Zone map is a marvelously useful tool for gardeners.

In this book I stick with the 1990 release of the USDA Growing Zone Map, as the 2012 version represents many areas as too warm.

My old neighborhood in North Florida is now solidly in "zone 9" on the new map—which is crazy, as that would make it possible for me to grow starfruit there, and that is most definitely not possible—and much of the nation has been declared warmer even as some areas face record lows. The warming is perhaps due to the heat island effect, a lurching forward and backwards of the climate that skews the data, or more sinisterly, political machinations related to the theory of Global Warming (though, to their credit, the USDA has denied a link). Whatever the reasoning or your thinking on climate change, you are safer sticking to the map from 1990, as being too cheery about the prospects for your tropical plants will only lead to pain, which leads to anger which leads to suffering, and that is the path to the Dark Side. Or something like that.

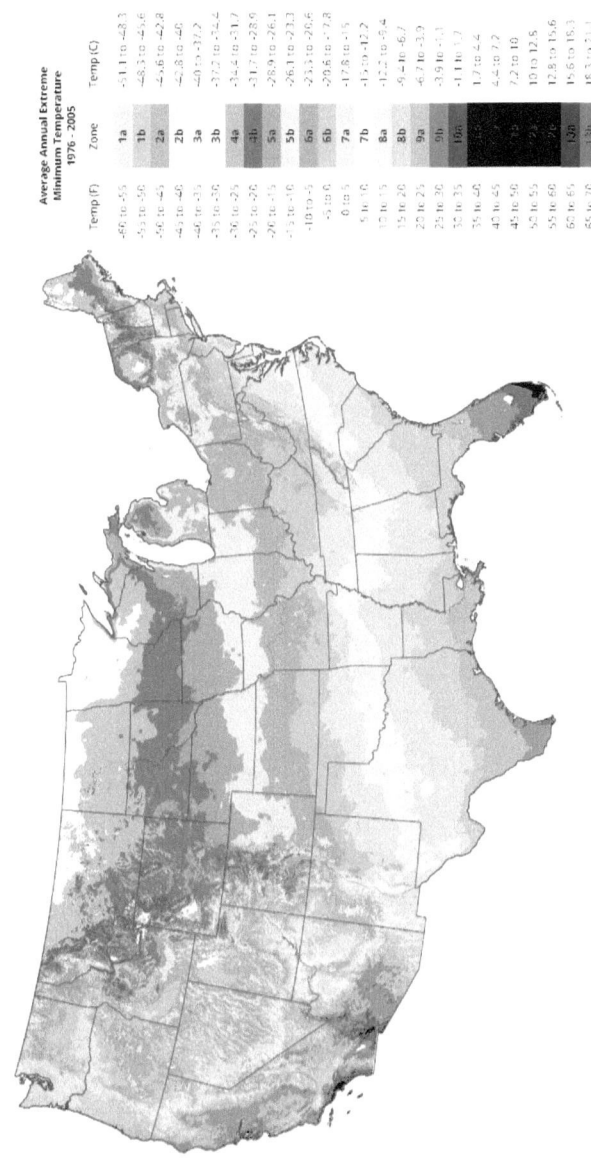

Appendix B: Coconuts Outside the Tropics?

If you live in a mild climate and really have too much time on your hands, why not consider growing some coconut palms in your backyard, far from their natural home on sunny, tropical shores?

If you drive South on I-95 through my home state of Florida, it's readily obvious where the coconut palms start. It's like you cross a line somewhere a little north of Palm Beach... and then there they are. Beautiful, tropical coconut palms.

Florida isn't a tropical state for the most part. It's close, but not quite. North of the southern tip of the state, with the exception of a few sheltered areas, the coconuts disappear.

On my North Florida homestead a bit south of Gainesville, it was impossible to grow coconut palms. The overnight lows would sometimes hit the teens. On one night, I measured a low of 12°F. That's 20 degrees below weather that can kill a coconut palm.

So I did some thinking about coconut palms and the way they grow. First, they need warmth year-round—or at least

protection from freezing weather, and they need space to reach maturity.

A greenhouse is the logical option for growing coconuts outside the tropics, right?

Sure... but do you have any idea how tall coconut palms can get?

The full-sized varieties can almost hit 100 feet, so that's obviously not going to work.

Fortunately, there are dwarf varieties of coconut palms that will bear fruit at around 8 feet tall and only get perhaps to 20 feet.

"But wait," you say, "that's TOO tall for a backyard greenhouse!"

Correct. My favorite inexpensive greenhouse has 9 feet of clearance in the middle. Let's just say we're growing in one that has an 8 foot high roof. That's pretty normal for a backyard greenhouse.

So, as already discussed, the only place to go is down.

Dig a pit beneath the greenhouse and put the greenhouse over the top of it. This will gain you another 6 feet of headspace.

"But wait... that's just a little baby coconut palm in the picture you drew, David The Good! And the complete height is only 14 feet, which is better than 8 feet... but still, you said

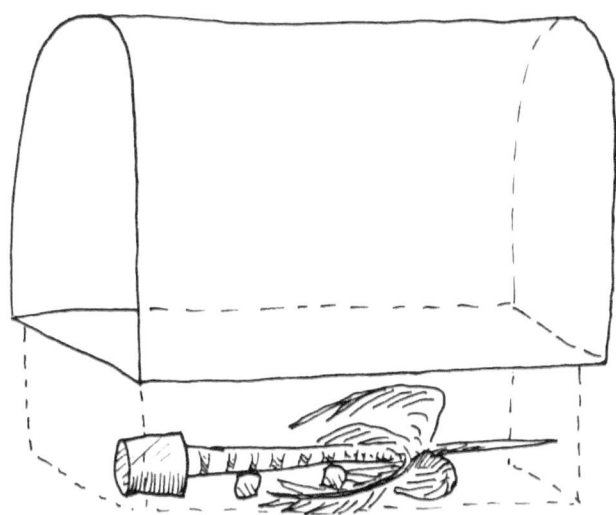

the coconut palm could get up to 20 feet tall!"

Very good objection, but think about it further. Have you ever seen pictures of coconut palms growing sideways along the shoreline?

Coconuts have the ability to be blown sideways, then recover, bend upwards and keep growing.

What if we used that ability to our advantage?

What if one planted a coconut palm in a large pot at one end of the greenhouse, then let it grow until it almost hit the ceiling, then tipped it over.

"No! It looks dead!"

It'll be fine, don't worry. It looks sad now, but wait!

After laying the coconut palm on its side, I'd also put

a couple of blocks beneath the trunk to support it, which also would make it into a sitting bench for one side of your greenhouse. In a short period of time, the coconut tree is going to start growing upwards again.

Now your greenhouse coconut palm tree can hit a productive size without knocking the top off its shelter... and you could potentially produce your own coconuts far from the sunny tropics.

Further Thoughts and Possibilities

Since I haven't tested this idea, I can't say for sure if it will work. My gut says it will, judging by what I've seen of coconut palms as well as the fact that I've seen pit greenhouses like the one in the drawings.

Adding barrels of water would be a good idea since coconuts really, really don't like cold.

Other possibilities for growing palms (though not necessarily coconut palms) outside the tropics and for growing tropical plants in regions where they're not "supposed" to grow are discussed in the fascinating book *Palms Won't Grow Here and Other Myths*, by David A. Francko.

Though the book is focused on ornamental species, the ideas Dr. Francko unlocks are worth the price of admission. The man has done amazing things and it's fun to hear about just how crazy you can get with pushing growing zones.

So... anyone up for the challenge of growing coconut palms

in a non-tropical climate? Pick up the tiki torch and make it happen!

Endnotes

1. See https://epa.gov/heat-islands.

2. britannica.com/topic/smudge-pot
3. G. Carroll Rice, 2013, "Those Damn Smudge Pots!", El Cajon Historical Society.
4. The Vagabond, "Light The Pots", *The Harvard Crimson*.

5. gardenpool.org
6. David Francko, *Palms Won't Grow Here and Other Myths*, pg. 37

7. Messy Nessy, "The Last Peach Orchards of Paris" messynessychic.com/2014/03/04/the-last-peach-orchards-of-paris/.

8. crfg.org/pubs/ff/jaboticaba.html
9. edis.ifas.ufl.edu/pdffiles/Mg/Mg37000.pdf
10. *Agricultural Options for Small-Scale Farmers,* echonet.org, p. 202.

www.ingramcontent.com/pod-product-compliance
Lightning Source LLC
Chambersburg PA
CBHW021426070526
44577CB00001B/74